Linux for System Administrators

Navigate the complex landscape of the Linux OS and command line for effective administration

Viorel Rudareanu

Daniil Baturin

BIRMINGHAM—MUMBAI

Linux for System Administrators

Copyright © 2023 Packt Publishing

Group Product Manager: Pavan Ramchandani

Publishing Product Manager: Neha Sharma

Senior Editor: Arun Nadar

Technical Editor: Yash Bhanushali

Copy Editor: Safis Editing

Project Coordinator: Ashwin Kharwa

Proofreader: Safis Editing

Indexer: Pratik Shirodkar

Production Designer: Vijay Kamble

Marketing Coordinator: Marylou De Mello

First published: 1250823

Production reference: 1280823

Published by Packt Publishing Ltd.
Grosvenor House
11 St Paul's Square
Birmingham
B3 1RB

ISBN: 978-1-80324-794-6

www.packtpub.com

Contributors

About the authors

Viorel Rudareanu has over 15 years of experience working with Linux. He possesses a comprehensive understanding of operating systems and their intricacies. He has successfully implemented and managed Linux-based infrastructures, leveraging the platform's robustness and flexibility to optimize system performances.

I want to thank the people who have been close to me and supported me, especially my wife, Janina, and my two daughters, who sometimes missed me while writing the book.

Daniil Baturin was first introduced to Linux by older friends when he was a high school student in 2002, and he has worked with Linux systems professionally since 2007 – as a systems administrator, software engineer, and contributor to multiple open source projects. He's currently a maintainer of VyOS – a Linux-based network operating system focused on enterprise and service provider routers.

I would like to say thanks to all people who work on open source software.

About the reviewers

Himanshu Sharma has nearly 18 years of experience in designing, architecting, and developing cloud and network software. He has worked for some of the biggest companies, such as Brocade and Juniper, and start-ups, such as Netskope. Currently, he works with Netskope as a principal engineer, responsible for Netskope's security service offering. He designed, architected, and developed Netskope's advanced threat protection services from the ground up. He has a keen interest and experience in developing scalable cloud services with cutting-edge technologies. His favorite hobbies are skiing and playing video games.

I would like to thank my wife, Puja, who gave me all the support I needed, and my two loving and beautiful daughters, Navya and Jaanvi.

Also, I want to thank my brother Sudhanshu for always having my back and my parents for all their sacrifices to get me where I am today.

Lekshmi Narayan Kolappan is an exceptional trailblazer in the realm of SRE/DevOps engineering. With an illustrious decade-long career in the software industry, his expertise shines through in the areas of AWS DevOps, SRE, Kubernetes, Docker, and Linux.

He is currently a site reliability engineer at Extreme Reach Limited, London, with a focus on implementing cutting-edge observability tools and performance monitoring.

Beyond technical brilliance, his dedication to refining the **software development life cycle (SDLC)** is unparalleled. He streamlines processes, crafts infrastructure audits, and automates testing frameworks.

Beyond the corporate world, in his spare time, he delights in exploring photography, literature, and tech blogs.

I wish to extend my heartfelt gratitude to my wife and my little son, who wholeheartedly comprehend the time and dedication involved in my pursuit of growth and development in the open source world. Their unwavering support is the cornerstone of my journey, and I truly cherish their understanding and encouragement.

Table of Contents

4

Processes and Process Control   37

5

Hardware Discovery   51

Part 2: Configuring and Modifying Linux Systems

6

7

8

9

Network Configuration and Debugging   131

10

Storage Management   157

Part 3: Linux as a Part of a Larger System

11

12

13

14

Automation with Chef   235

Preface

Linux has become one of the most widely used operating systems in the world, powering everything from servers to smartphones. With its robust security, stability, and flexibility, Linux has become the go-to choice for many system administrators who seek a reliable and efficient platform to manage their organization's IT infrastructure.

As a system administrator, you are responsible for managing the day-to-day operations of your organization's IT infrastructure. This includes everything from setting up and configuring servers and maintaining network connectivity to troubleshooting issues when they arise. To do all of this effectively, you need a solid understanding of Linux and its various tools and utilities.

Throughout the book, you will find real-world examples and hands-on exercises that will help you build practical skills and gain confidence in your ability to manage Linux systems. You will also learn about the latest tools and techniques for managing large-scale Linux environments.

We hope that this book will serve as a valuable resource for you as you navigate the world of Linux system administration. Whether you are just starting out or looking to deepen your knowledge and skills, *Linux for System Administrators* will provide you with the knowledge and tools you need to succeed.

Who this book is for

Whether you are new to Linux or have been using it for years, this book provides a comprehensive overview of the operating system, its tools, and its best practices. The book covers everything from basic Linux concepts to more advanced topics, such as server virtualization, network configuration, and system security.

What this book covers

Chapter 1, *Getting to Know Linux*, provides an overview of the Linux operating system. It covers the basics of Linux's history, its features, and how it differs from other operating systems such as Windows and macOS. The goal of this chapter is to provide you with a foundational understanding of Linux and its key concepts so that you can use the operating system effectively.

Chapter 2, *The Shell and its Commands*, provides examples of how to use specific commands and options, and it also covers more advanced topics such as shell scripting, piping and redirection, and using regular expressions. The goal of this chapter is to provide you with a solid understanding of the shell and its basic commands, enabling you to efficiently navigate and manage your Linux or Unix-like systems from the command line.

Chapter 3, The Linux Filesystem, covers the structure and organization of the filesystem used by the system. The chapter begins with an overview of the filesystem hierarchy, including the `root` directory and its subdirectories, such as `/bin`, `/etc`, `/home`, `/usr`, and `/var`. It also covers the different types of files and directories found in the filesystem, including regular files, directories, symbolic links, and special files such as device files. It also discusses file permissions and ownership, and how to use commands such as `chmod` and `chown` to modify them.

Chapter 4, Processes and Process Control, begins with an overview of processes and their properties, including process IDs, parent process IDs, and process statuses. It then covers how to view and manage running processes, using tools such as `ps`, `top`, and `kill`. Understanding these concepts can help system administrators optimize system performance and troubleshoot issues related to process management.

Chapter 5, Hardware Discovery, gives an overview of the different types of hardware components found in a typical computer system, including processors, memory, storage devices, and input/output devices. It also covers how the operating system detects and identifies these components, using tools such as `dmesg`, `lspci`, and `lsusb`.

Chapter 6, Basic System Settings, covers the configuration of basic system settings that affect a system's behavior and performance. This chapter is essential for system administrators and users who need to customize the system to meet their specific requirements.

Chapter 7, User and Group Management, begins with an overview of user accounts and groups and their properties, including user IDs, group IDs, home directories, and shell settings. It then covers how to create and manage user accounts and groups, using tools such as `useradd`, `usermod`, `groupadd`, and `groupmod`.

Chapter 8, Software Installation and Package Repositories, starts with the installation and management of software packages on the system. This chapter is essential for system administrators and users who need to install, upgrade, and manage software packages to meet their requirements.

Chapter 9, Network Configuration and Troubleshooting, begins with an overview of network configuration and networking tools available on the system, such as `ifconfig`, `ip`, and `netstat`. It then covers how to configure network interfaces, assign IP addresses and netmasks, and configure network routing. This chapter is essential for system administrators and users who need to set up and maintain network connectivity and services.

Chapter 10, Storage Management, gives an overview of storage devices and filesystems and their properties, such as device names, device IDs, and mount points. It then covers how to create, manage, and mount filesystems, using tools such as `fdisk`, `mkfs`, and `mount`. Other topics covered include managing storage devices, such as partitioning and formatting disks, and managing **Logical Volume Manager** (**LVM**) volumes. Understanding these concepts and how to manage them is essential to ensure reliable and secure storage infrastructure. This chapter is essential for system administrators and users who need to manage storage resources, such as hard drives, solid-state drives, and network-attached storage.

Chapter 11, Logging Configuration and Remote Logging, includes configuring log forwarding and aggregation, setting up centralized logging servers, and analyzing system logs. Understanding these concepts and how to manage them is essential to ensure reliable and secure system logging infrastructure.

Chapter 12, Centralized Authentication, begins with an overview of authentication and authorization mechanisms available on the system, such as local password files, **Lightweight Directory Access Protocol** (**LDAP**), and Active Directory. It then covers how to configure and manage user authentication using tools such as **Pluggable Authentication Module** (**PAM**) and **Name Service Switch** (**NSS**). It also covers the configuration and management of user authentication and authorization on the system. This chapter is essential for system administrators who need to manage user access and privileges across multiple systems.

Chapter 13, High Availability, includes configuring and managing cluster resources, such as IP addresses, network interfaces, and shared storage devices; configuring and managing cluster services, such as web servers, databases, and email servers; and monitoring and troubleshooting cluster operations. Understanding these concepts and how to manage them is essential to ensure the high availability and reliability of critical applications and services.

Chapter 14, Automation with Chef, gives an overview of infrastructure automation and configuration management concepts, such as idempotence, the declarative approach, and the **Infrastructure as Code** (**IaC**) paradigm. It then covers how to use Chef to automate the configuration and management of systems, including nodes, cookbooks, recipes, and resources.

Chapter 15, Security Guidelines and Best Practices, covers how to implement security measures and best practices.

To get the most out of this book

To follow along the topics covered in this book, you just need a Linux VM or a Linux machine.

Conventions used

There are a number of text conventions used throughout this book.

`Code in text`: Indicates code words in text, database table names, folder names, filenames, file extensions, pathnames, dummy URLs, user input, and Twitter handles. Here is an example: "For example, the configuration file directory for the Apache HTTP server is `/etc/httpd` on Red Hat Linux derivatives, but `/etc/apache2` on Debian derivatives."

A block of code is set as follows:

```
global_defs {
    notification_email {
        admin@example.com
        webmaster@example.com
    }
    notification_email_from keepalived@example.com
    smtp_server 203.0.113.100
    smtp_connect_timeout 30
}
```

Any command-line input or output is written as follows:

```
$ echo '#!/bin/bash' >> hello.sh
$ echo 'echo "hello world"' >> hello.sh
$ chmod +x ./hello.sh
$ ./hello.sh
hello world
```

Bold: Indicates a new term, an important word, or words that you see on screen. For instance, words in menus or dialog boxes appear in **bold**. Here is an example: "The other is **Red Hat Package Manager (RPM)**, which is used with the rpm utility and is developed by Red Hat."

> **Tips or important notes**
> Appear like this.

Get in touch

Feedback from our readers is always welcome.

General feedback: If you have questions about any aspect of this book, email us at customercare@packtpub.com and mention the book title in the subject of your message.

Errata: Although we have taken every care to ensure the accuracy of our content, mistakes do happen. If you have found a mistake in this book, we would be grateful if you would report this to us. Please visit www.packtpub.com/support/errata and fill in the form.

Piracy: If you come across any illegal copies of our works in any form on the internet, we would be grateful if you would provide us with the location address or website name. Please contact us at copyright@packtpub.com with a link to the material.

If you are interested in becoming an author: If there is a topic that you have expertise in and you are interested in either writing or contributing to a book, please visit authors.packtpub.com.

Share Your Thoughts

Once you've read *Linux for System Administrators*, we'd love to hear your thoughts! Scan the QR code below to go straight to the Amazon review page for this book and share your feedback.

https://packt.link/r/1803247940

Your review is important to us and the tech community and will help us make sure we're delivering excellent quality content.

Download a free PDF copy of this book

Thanks for purchasing this book!

Do you like to read on the go but are unable to carry your print books everywhere?

Is your eBook purchase not compatible with the device of your choice?

Don't worry, now with every Packt book you get a DRM-free PDF version of that book at no cost.

Read anywhere, any place, on any device. Search, copy, and paste code from your favorite technical books directly into your application.

The perks don't stop there, you can get exclusive access to discounts, newsletters, and great free content in your inbox daily

Follow these simple steps to get the benefits:

1. Scan the QR code or visit the link below

https://packt.link/free-ebook/9781803247946

2. Submit your proof of purchase
3. That's it! We'll send your free PDF and other benefits to your email directly

Part 1:
Linux Basics

The highest priority for a beginner systems administrator is to learn what the operating system consists of and how to interact with it. In the first part of this book, you will learn a brief history of Linux, how Linux-based systems are used in the real world, and what the relationship between the Linux kernel and the wider open source software ecosystem is. By the end of this part, you will be able to use basic commands to navigate the system and manage files, processes, and hardware devices.

This part has the following chapters:

- *Chapter 1, Getting to Know Linux*
- *Chapter 2, The Shell and its Commands*
- *Chapter 3, The Linux Filesystem*
- *Chapter 4, Processes and Process Control*
- *Chapter 5, Hardware Discovery*

1
Getting to Know Linux

Linux is a family of operating systems based on the same kernel. Since it's a family of independently developed systems that have different design principles, goals, and implementation details, it's important to understand what makes that very situation possible and how those systems are structured. In this chapter, we will discuss the concept of Linux distributions and open source software licensing and see how Linux-based systems are used. We will cover the following topics:

- The structure of a Linux system
- Open source software licenses
- Linux usage in the modern world

The structure of a Linux system

Linux and its multiple distributions often seem complicated for beginners. To clarify this, let's examine the structure and evolution of operating systems in general.

The Linux kernel and Linux-based operating systems

When people say *Linux*, they may mean different things. In the narrow sense, Linux is an operating system kernel that was created in the early 90s by Linus Torvalds and is now developed and maintained by a large international community. However, when people say they are using Linux, they usually mean a family of operating systems that use that kernel and usually (but not always) a set of system libraries and utilities created by the GNU project, which is why some insist that such systems should be referred to as GNU/Linux instead.

> **Note**
>
> The GNU project is a free software project that was launched in 1983 by Richard Stallman. His goal was to create a complete Unix-like operating system composed entirely of free software. **GNU** stands for **GNU's Not Unix**, which reflects the project's goal of creating a free software alternative to the proprietary Unix operating system.

To fully understand how that unusual situation became possible, let's briefly discuss the history of operating systems.

Kernel versus user space

The earliest computers had very low computational power, so they would only have one program in their memory at a time, and that program had complete control over the hardware. As computing power increased, it became feasible to have multiple users use the same computer at the same time and run multiple programs – an idea known as time-sharing or multitasking. Shared computers would run a program known as a supervisor that would allocate resources to end user programs. A set of supervisor programs and system utilities became known as an operating system. The earliest time-sharing systems used cooperative multitasking, where programs were expected to transfer control back to the supervisor on their own. However, if a programming mistake made a program run into an endless loop or write data to a wrong memory address, such a program could cause the entire computer to hang or corrupt the memory of another program, including the supervisor.

To make multitasking more reliable, newer generations of hardware introduced protection mechanisms that allowed a supervisor program to take control of the CPU back from end user programs and forcibly terminate programs that tried to write something to memory that belonged to other programs or the supervisor itself.

That brought a separation between the operating system kernel and user space programs. End user programs physically couldn't control the hardware directly anymore, and neither could they access memory that wasn't explicitly allocated to them. Those privileges were reserved for the kernel – the code that includes a process scheduler (serving the same purpose as old supervisor programs) and device drivers.

Inside a single program, programmers are free to organize their code as they see fit. However, when multiple independently developed components need to work together, there needs to be a well-defined interface between them. Since no one writes directly in machine code anymore, for modern systems, this means two interfaces: the **Application Programming Interface** (**API**) and the **Application Binary Interface** (**ABI**). The API is for programmers who write source code and define function names they can call and parameter lists for those functions. After compilation, such function calls are translated into executable code that loads parameters into the correct places in memory and transfers control to the code to be called – where to load those parameters and how to transfer control is defined by the ABI.

Interfaces between user space programs and libraries are heavily influenced by the programming language they are written in.

On the contrary, interfaces between kernels and user space programs look more similar to hardware interfaces. They are completely independent of the programming language and use software interrupts or dedicated *system call* CPU instructions rather than the function calls familiar to application programmers.

> **Note**
>
> A system call in Linux is a mechanism that allows user-level processes to request services from the kernel, which is the core of the operating system. These services can access hardware devices, manage processes and threads, allocate memory, and perform other low-level tasks that require privileged access.

Those interfaces are also very low-level: for example, if you want to use the `write()` system call to print a string to standard output, you must always specify how many bytes to write – it has no concept of a *string variable* or a convention for determining its length.

For this reason, operating systems include standard libraries for one or more programming languages, which provide an abstraction layer and a stable API for end user programs.

Most operating systems have the kernel, the standard libraries for programming languages, and often the basic system utilities developed by a single group of people in close collaboration, and all those components are versioned and distributed together. In that case, the kernel interface is usually treated as purely internal and isn't guaranteed to remain stable.

The Linux kernel and the GNU project

Linux is unique in that it was developed to provide a replacement kernel for an existing user space part of an operating system. Linus Torvalds, the founder of the project, originally developed it to improve the functionality of MINIX – an intentionally simplified Unix-like operating system meant for instruction rather than production use. He's since been using the GNU C compiler and user space programs from the GNU project – the project that Richard Stallman started with the goal to create a complete Unix-like operating system that would be free (as in freedom) and open source, and thus available for everyone to use, improve, and redistribute.

At the time, the GNU project had all the user space parts of an operating system, but not a usable kernel. There were other open source Unix projects, but they were derived from the BSD Unix code base, and in the early 90s, they were targets of lawsuits for alleged copyright infringement. The Linux kernel came at a perfect time since Linus Torvalds and various contributors developed it completely independently and published it under the same license as the GNU project software – the GNU **General Public License (GPL)**. Due to this, a set of GNU software packages, plus the Linux kernel, became a possible basis for a completely open source operating system.

However, Linus Torvalds wasn't a GNU project member, and the Linux kernel remained independent from the **Free Software Foundation (FSF)** – it just used a license that the FSF developed for the GNU project, but that any other person could also use, and many did.

Thus, to keep new Linux kernel versions useful together with the GNU C library and software that relied on that library, developers had to keep the kernel interface stable.

The GNU C library wasn't developed to work with a specific kernel either – when that project started, there wasn't a working GNU kernel, and GNU software was usually run on other Unix-like operating systems.

As a result, both Linux and the GNU software can be and still are used together and in different combinations. The GNU user space software set can also be used with the still-experimental GNU hard kernel, and other operating systems use it as system or add-on software. For example, Apple macOS used GNU Bash as its system shell for a long time, until it was replaced by zsh.

The stability guarantees of the Linux kernel interface make it attractive to use as a basis for custom operating systems that may be nothing like Unix – some of them just have a single program run on top of the kernel. People have also created alternative standard libraries for different programming languages to use with Linux, such as Musl and Bionic for the C programming language, which use more permissive licenses and facilitate static linking. But to understand those licensing differences, we need to discuss the concept of software licenses.

Open source software licenses

A software license is an agreement between a copyright holder and a recipient of the software. Modern copyright laws are designed to give authors complete control over the use and distribution of their work – copyright automatically exists from the moment a piece of work is fixed on any medium and no one can use or copy that work without explicit permission from the author. Thus, a license agreement is required to grant a user some of the permissions that are reserved for the author by default. Authors are free to specify any conditions, and many individuals and companies use that to restrict what users can do – for example, only permit non-commercial use. A license agreement is also normally made between an author or a copyright holder and a specific person.

However, in the late 1980s, programmers and lawyers came up with the idea to use authors' unlimited control over their works to ensure that anyone can use, distribute, and modify them rather than prevent that. They introduced *public licenses*, which grant permissions to everyone rather than just people who signed or otherwise accepted the agreement, and wrote several reusable license agreements that anyone could apply to their software. That concept became known as **copyleft** – a reversal of copyright. Those licenses became known as open source licenses because they explicitly permit the distribution and modification of software source code.

All the *classic* licenses were born in that period: the MIT license, the BSD license, the GNU GPL, and the GNU **Lesser/Library General Public License** (**LGPL**). None of those licenses limit users' rights to use software distributed under them. Conditions, if any, apply only to the distribution and modification of executables and source code.

Permissive and copyleft licenses

When it comes to distribution, two schools of thought differ in their approach to distribution conditions.

Proponents of permissive licenses believe that recipients of software must have absolute freedom to do anything with it, even to incorporate it into other software that isn't open source or to create closed source derivatives. The MIT and BSD licenses are typical examples of permissive open source licenses.

Proponents of copyleft believe that it's important to protect open source software from attempts to appropriate the work of its authors and create a closed source derivative. The GNU GPL is the purest example of this – if anyone distributes executables of programs under the GPL or programs that link with libraries under the GPL, they must also distribute the source code of that program under that license. This is the most radical approach and is known as *strong copyleft*.

Licenses that allow you to link libraries to programs under any other license but require library code modifications to be under the same license are known as *weak copyleft* licenses. The most widely used example is the GNU LGPL.

Patent grant, tivoization, and SaaS concerns

The GNU GPL was created as a response to the rise of proprietary software distributed without source code, which prevented end users from improving it and sharing their improvements. However, the software industry is evolving, and new trends are appearing that some see as threats to the existence or financial sustainability of open source software.

One such threat is *patent trolling* – the use of software patents (in jurisdictions where they exist) in bad faith. As a response to it, some newer licenses and new versions of old licenses, such as the Apache license and the GNU GPLv3, introduced a patent grant clause. Such a clause prevents contributors to the source code of software from making patent claims against its users. If they make such legal threats, their licenses are revoked.

A more controversial point of the GPLv3 is its attempts to protect users' rights to run modified versions on their hardware. The practice of preventing hardware from running custom software through digital signatures and similar mechanisms is sometimes called *tivoization*, after a Linux-based digital video recorder named TiVo that was an early example of such lock-in. While some projects supported the idea to prevent it, for others, the GPLv3 clause was a reason not to switch from GPLv2 – the Linux kernel is among those projects that stayed at the old GPL version.

Finally, all classic licenses were written in a time when all software was deployed on-premises, while in the modern world, a lot of software is delivered over a network and its executables aren't accessible to end users – an approach known as **Software-as-a-Service** (**SaaS**). Since the GPL says that every recipient of a binary executable is entitled to receive its source code, it does not apply to SaaS since the user never receives any executables. This allows vendors to create modified versions of the software under the GPL without sharing their improvements with the community. Several licenses were developed in response to that trend, such as the Affero GPL.

In the last few years, big technology companies that provide hosted versions of open source software started to be seen as undermining project maintainers' ability to earn money from services since it's very difficult to compete on price with effective monopolies. In response, some projects started switching to licenses that have restrictions on usage, which many argue are no longer open source licenses, even though the source code is still available. The future of such licenses is an open question.

Linux distributions

The fact that software under open source licenses is free to modify and distribute made it possible to assemble complete operating systems with kernels, system libraries, and utilities, as well as a selection of application software. Since open source licenses have no restrictions on usage, there is no need to make the user accept a license agreement for each component.

In the early days of Linux, setting up a usable Linux environment was a complicated and tedious endeavor. To make that process simpler, Linux enthusiasts started preparing the first *distributions* – sets of packages and scripts to automate their installation. Many of those early distributions, such as Softlanding Linux System and Yggdrasil, are now defunct, but some are still maintained – Slackware Linux is a prominent example.

Package managers and package repositories

Early distributions had a relatively humble goal, which was to provide users with a working barebones system that they could then install their application software on. However, later distributions set out to rethink the process of software installations. The number of open source software projects was growing, and CD drives and internet connections were also becoming more affordable, so it was feasible to include much more software in a distribution than ever before.

However, many applications depend on shared libraries or other applications. Traditionally, installation packages would either include all dependencies or leave that dependency management to the user. Since distribution is managed by a single group of maintainers, developers came up with the idea of sharing dependencies between all packages that need them and automatically installing all dependencies when a user requested the installation of a package. That gave rise to package managers and package repositories – collections of files in a special format, including compiled binaries and metadata such as the package version and its dependencies.

The two most popular package formats and package managers that work with them were developed in the mid-90s and are still in use today. One is the DEB format, which is used with the dpkg utility, developed by Debian. The other is **Red Hat Package Manager** (**RPM**), which is used with the rpm utility and is developed by Red Hat.

The dpkg and rpm tools are responsible for installing package files on local machines. To install a package, the user needs to retrieve the package itself and all packages it depends on. To automate that process, distributions developed high-level package managers that can automatically download packages from online repositories, check for updates, search metadata, and more. Those high-level package managers usually rely on low-level ones to manage the installation. Debian's **Advanced Packaging Tool** (**APT**) usually works with DEB packages, although it's technically possible to use it with RPM. High-level package managers that primarily use the RPM format are more numerous: YUM and DNF, which are maintained by Red Hat, zypper from openSUSE, and urpmi, which is developed for the now-defunct Mandrake Linux and still used by its forks.

Many of the currently existing distributions have either been actively maintained since the 90s or are forks that split off at different points in time. For example, Ubuntu Linux was forked from Debian GNU/Linux in the early 2000s, while Rocky Linux is a Red Hat Enterprise Linux derivative that started in 2021.

However, completely independent distributions also appear once in a while. Some of them are special-purpose systems that have requirements that classic general-purpose distributions cannot fulfill. For example, OpenWrt is a Linux-based system for consumer routers, originally developed for the Linksys WRT-54G device, hence the name. Such devices often have just a few megabytes of flash drive space, so operating systems for them have to be very compact, and they also have to use special filesystems such as JFFS that are designed for NAND flash drives.

Other independent distributions experiment with different package management and installation principles. For example, NixOS and GNU Guix use an approach that allows the user to revert system updates if anything goes wrong with new package versions.

In this book, we will focus on Debian/Ubuntu and Red Hat-based systems because they have been the most popular distributions for a long time and remain popular.

Differences between distributions

The differences between distributions do not stop at package managers. Configuration file locations may differ, and default configurations for the same packages may also differ dramatically. For example, the configuration file directory for the Apache HTTP server is `/etc/httpd` on Red Hat Linux derivatives, but `/etc/apache2` on Debian derivatives.

Some distributions also use high-level configuration tools and you may take them into account.

The choice of software and its ease of installation may also differ. Debian, Fedora, and many other distributions leave the choice of a desktop environment to the user and make it easy to install multiple different desktop environments on the same system so that you can switch between GNOME3, KDE, MATE, or anything else for different login sessions. In contrast, the Ubuntu family of distributions includes multiple flavors for different desktop environments and expects that if you don't like its default choice (the Unity desktop environment), you should use Kubuntu for KDE, for example, rather than the default Ubuntu. Finally, some distributions come with a custom desktop environment and don't support anything else, such as elementary OS.

However, experienced Linux users can usually find their way around any distribution.

Linux usage in the modern world

The open source nature of the Linux kernel and its support for multiple hardware architectures made it a very popular choice for custom operating systems, while general-purpose Linux distributions also found wide use in every niche where proprietary Unix systems were used before.

The most popular Linux-based operating system in the world is Android. While most Android applications are written for a custom runtime and never use any functionality of the Linux kernel directly, it's still a Linux distribution.

Network devices are usually managed through a web GUI or a custom command-line interface, but they still often have Linux under the hood. This applies to consumer-grade Wi-Fi routers, as well as high-performance enterprise and data center routers and switches alike.

General-purpose Linux distributions are also everywhere. Thanks to built-in support for running virtual machines (through Xen or KVM hypervisors), Linux powers the largest cloud computing platforms, including Amazon EC2, Google Cloud Platform, and DigitalOcean. A lot of guest systems are also Linux machines running web servers, database systems, and many other applications.

Linux is also widely used in high-performance computing: all of the most powerful supercomputers in the world are now running Linux on their control and I/O nodes.

Last but not least, the author of this chapter typed these words on a Linux desktop.

Summary

A Linux distribution is a complete operating system that includes the Linux kernel and a set of libraries and programs developed by various people and companies. The Linux kernel and core system libraries are not developed by the same group. Instead, the Linux kernel provides a stable ABI that allows anyone to develop a standard library for a programming language to run on top of it.

Open source licenses come with different conditions, but they all allow anyone to use the software for any purpose and distribute its copies, and that's what makes the existence of Linux distributions possible.

Different distributions use different approaches to package management and configuration, but experienced users can learn how to use a new distribution fairly quickly if they know the fundamentals. These days, Linux can be found everywhere, from mobile phones to the most powerful supercomputers.

In the next chapter, we will learn about the various shells available on Linux systems, as well as basic commands.

2
The Shell and Its Commands

We'll be doing a lot of things in the shell, such as installing packages, making new users, creating directories, modifying permissions of files, and so on. These will be the basics but will be your first interaction with the shell to understand what is happening behind it and to get more confident. In order to improve our effectiveness with the shell, we'll be devoting an entire chapter to it this time around.

In this chapter, we will cover the following topics:

- A basic definition of a shell in order to understand how it works, including an overview of its features and a description of the most common shells

- How to use basic commands in order to get familiarized with Linux (in this chapter, CentOS version 8 will be used)

- Basic notions about how to use commands to change the ownership of files and directories

What is a shell?

Computer software known as a *shell* makes an operating system's services accessible to users or other programs.

A shell is a program that receives commands and sends them to the operating system for processing, to put it simply. In an interactive session, the user has the option of typing commands from the keyboard, or they can be written in a shell script that can be reused. On a Unix-type system such as Linux in the past, it was the sole **user interface** (**UI**) accessible. Today, in addition to **command-line interfaces** (**CLIs**) such as shells, we also have **graphical UIs** (**GUIs**).

The fundamental capability of shells is the ability to launch command-line programs that are already installed on the system. They also offer built-ins and scripting control structures such as conditionals and loops. Each shell has its own way of doing that. Some shells still support the Bourne shell, one of the original shells that was created for an early Unix system by a programmer named Steve Bourne and later standardized in the **Portable Operating System Interface** (**POSIX**) standard. Other projects, such as csh/tcsh, zsh, and fish, purposefully utilize a different syntax.

In order to use command-line shells, a user must be knowledgeable about commands, their calling syntax, and the fundamentals of the shell's specific scripting language.

A Linux user can utilize a variety of shells, including the following:

- `sh`: A POSIX-compatible Bourne shell. In modern distros, it's usually just **Bourne again shell** (**Bash**) running in compatibility mode.

- `csh/tcsh`: These come from the **Berkeley Software Distribution (BSD)** Unix system family but are also available on Linux; their scripting syntax is similar to that of C, and they are incompatible with the Bourne shell.

- `ksh`: A Bourne shell derivative that was once very popular.

- `bash`: Bash is the most common Linux shell and was created for the GNU project.

- `zsh` and `fish`: These are highly customizable and feature-rich shells that are intentionally different from `sh` derivatives and require learning, but have large communities of enthusiasts.

They all share similar properties, but each has its own unique attributes.

In this book, we will assume you are using Bash since it's the default in most Linux distributions.

The Unix shell and Bash command language were both developed by Brian Fox for the GNU project. These were intended to serve as free software replacements for the Bourne shell. Since its introduction in 1989, it has remained the default login shell for the vast majority of Linux distributions. Linus Torvalds ported Bash and the **GNU Compiler Collection (GCC)** to Linux as one of the initial applications.

Bash has the following features:

- The shell will check to see whether a command is built in before searching through a list of directories to locate the program if not. This set is known as the search path. By running the `echo $PATH` command in Bash, you can view it. The home directory and its subdirectory are included in the search path in addition to the current directory. You are able to create your own programs and call them up just by inputting their names. No matter which directory you are now in, a program such as this will be found and launched if it is stored in the `bin` directory. We will find out more about the Linux directory structure in *Chapter 3, The Linux Filesystem*.

- As with other Linux programs, the shell has a current directory linked to it. When looking for files, Linux-based programs begin in the current directory. To move the current directory to another location in the Linux filesystem, use the `cd` shell command. The current working directory is typically visible in the command prompt of modern shells. To check the version of your shell, run the `echo $SHELL` command. You will get an output such as `/bin/bash`.

- A command is executed by designating it. The majority of Linux commands are just programs that the shell runs. For instance, the following `ls` command scans the current directory and lists the names of its files: `ls -la`.

- Commands frequently have argument strings that could, for example, be filenames. For instance, the following command switches to the `tmp` directory in your `home` directory. The shell interprets the tilde character as your `home` directory:

  ```
  cd ~/tmp
  ```

- Multiple arguments are required for some commands. The copy command, for instance, requires two arguments: the file to copy and its destination. This is demonstrated as follows by copying `file1` to a new file, `file2`:

  ```
  cp file1 file2
  ```

- The flag or option argument strings for some commands typically start with `-`. The flags change how the invoked application behaves. When the following command is used, `ls` outputs a lengthy listing of files arranged by creation time:

  ```
  ls -lt
  ```

- Wildcards will be expanded by the shell to match filenames in the current directory. For example, to display a directory listing of files named `anything.sh`, type the following:

  ```
  ls -l *.sh
  ```

- **Standard input** (**stdin**) and **standard output** (**stdout**) are concepts that are followed by the majority of Linux commands and programs. The program receives a stream of data as stdin and produces a stream of output as stdout. These are frequently both connected to Terminal so that input is made via the keyboard and output is displayed on the screen. You can reroute stdin and stdout using the shell. `cat` is an abbreviation for *concatenate*. When run, the following command will show you the contents of one or more files without requiring you to open them for editing:

  ```
  cat /etc/passwd
  ```

- The shell has the ability to pipe data from one program's output to another's input. | is the pipe symbol. To count the number of words in `testfile.txt`, we can concatenate the file and pass the output into the `wc` program, like so:

  ```
  cat testfile.txt | wc -w
  1198
  ```

  Or, to count the number of lines from a `testfile.txt` file, we can use the following command:

  ```
  cat testfile.txt | wc -l
  289
  ```

- You can create aliases for commands or groups of commands that you use frequently or find difficult to input. For instance, we could use the `top10` alias to find the top 10 files in the current directory. `head` will show only the top lines. An alias is a shortcut for a command—for example, rather than remembering a very long command, you can create an alias that you can remember easily. Here's an example:

```
alias top10="du -hsx * | sort -rh | head -10"
```

- Some variables are predefined, such as $HOME, which is your `home` directory. To see a list of assigned variables, type the following command:

```
set
```

- A **manual** (**man**) page is like a manual with instructions and descriptions about each command. Run the following command to view the man page for Bash:

```
bash-3.2$ man bash
```

- *Scripts* of shell commands can be written. These can be called just like compiled programs (that is, just by naming them). For instance, we first create a file in `/bin` containing the following in order to construct a script named `top10.sh` that displays the top 10 biggest files in the current directory:

```
#! /bin/bash
du -hsx * | sort -rh | head -10
```

- We must next use the `chmod` command to make the file executable before we can run it normally:

```
chmod +x ~/bin/top10.sh
./top10.sh
```

See the man page on `bash` for more details (type `man bash`).

The up arrow key on the keyboard in Bash's extra mechanism enables you to access and modify past commands. The most recent command is displayed on Terminal again when you press the up arrow key. To access previous commands, press the up arrow key once more. Press *Enter* to run the command once more. Use the *Delete* key to remove characters from the command's end, or the back arrow key to move the cursor and change the command's contents by inserting or deleting characters.

By using the `history` command, you can view the history of commands.

You can rerun any command from the history by pressing ! and the line number—for example, !345.

Now that you know how to interact with the shell and what is happening when you type these commands, in the next section, we will try to practice some basic commands to make you more confident when you interact with Terminal.

Basic shell commands

Here's a rundown of some of the possible commands. For more information, see the man page for each command. Using the man command, you can view these online. Simply type man followed by the command name you wish to see (for example, if you want to learn more about the cat command, simply type man cat):

- pwd: The pwd command can be used to determine which directory you are in. Its name is an abbreviation for **print working directory**. It provides us with the absolute path, which is the path that begins at the root. The root directory is the foundation of the Linux filesystem. It's indicated by a forward slash (/). You can see the pwd command in use in the following screenshot:

```
[[voxsteel@centos8 ~]$ pwd
/home/voxsteel
[voxsteel@centos8 ~]$
```

Figure 2.1 – pwd command, showing the working directory

- mkdir: mkdir is the command to use when you need to make a new directory. Put mkdir packt on your command line to make a directory with that name. To list your created directory, use the ls -ld <directory_name> command. You can see the mkdir command in use here:

```
[[voxsteel@centos8 packt]$ mkdir new_directory
[[voxsteel@centos8 packt]$ ls -ld new_directory/
drwxrwxr-x. 2 voxsteel voxsteel 6 Mar 20 14:17 new_directory/
[voxsteel@centos8 packt]$
```

Figure 2.2 – mkdir command

- rmdir: To delete a directory, use rmdir. However, rmdir can only be used to remove an empty directory. To remove files and directories, use rm -rf directoryname/ (where -rf will recursively remove all the files and directories from inside the directory). To check whether a directory has been removed, use ls -ld <directory_name>. The rmdir command is shown here:

```
[voxsteel@centos8 packt]$ rmdir new_directory/
[voxsteel@centos8 packt]$ ls -ld new_directory
ls: cannot access 'new_directory': No such file or directory
[voxsteel@centos8 packt]$ 
```

Figure 2.3 – rmdir command

- `touch`: The command's initial intent was to set the file modification date to the current time. But since it will make a file if one doesn't already exist, it's frequently used to make empty files. Here's an example:

```
touch filename.txt
```

- `ls`: Use the `ls` command to see all files and directories inside the directory you are in. If you want to see hidden files, use the `ls -a` command. By using the `ls -la` command, you can see all the files and directories as a list, as illustrated here:

```
[[voxsteel@centos8 packt]$ ls
file1.txt  file2.txt  file3.txt  file4.txt  testfolder
[[voxsteel@centos8 packt]$ ls -a
.  ..  file1.txt  file2.txt  file3.txt  file4.txt  testfolder
[[voxsteel@centos8 packt]$ ls -la
total 4
drwxrwxr-x.  3 voxsteel voxsteel   92 Oct  8 17:46 .
drwx------. 18 voxsteel voxsteel 4096 Oct  8 17:45 ..
-rw-rw-r--.  1 voxsteel voxsteel    0 Oct  8 17:46 file1.txt
-rw-rw-r--.  1 voxsteel voxsteel    0 Oct  8 17:46 file2.txt
-rw-rw-r--.  1 voxsteel voxsteel    0 Oct  8 17:46 file3.txt
-rw-rw-r--.  1 voxsteel voxsteel    0 Oct  8 17:46 file4.txt
drwxrwxr-x.  2 voxsteel voxsteel    6 Oct  8 17:45 testfolder
[voxsteel@centos8 packt]$
```

Figure 2.4 – ls command

- `cp`: To copy files from the command line, use the `cp` command. This requires two arguments: the first specifies the location of the file to be copied, and the second specifies where to copy it. It could be a new folder or a new file (in case you need a copy of it). You can see the `cp` command in use here:

```
[[voxsteel@centos8 packt]$ cp file1.txt testfolder/
[[voxsteel@centos8 packt]$ cp file1.txt file5.txt
[[voxsteel@centos8 packt]$ ls -la
total 4
drwxrwxr-x.  3 voxsteel voxsteel  109 Oct  8 17:48 .
drwx------. 18 voxsteel voxsteel 4096 Oct  8 17:45 ..
-rw-rw-r--.  1 voxsteel voxsteel    0 Oct  8 17:46 file1.txt
-rw-rw-r--.  1 voxsteel voxsteel    0 Oct  8 17:46 file2.txt
-rw-rw-r--.  1 voxsteel voxsteel    0 Oct  8 17:46 file3.txt
-rw-rw-r--.  1 voxsteel voxsteel    0 Oct  8 17:46 file4.txt
-rw-rw-r--.  1 voxsteel voxsteel    0 Oct  8 17:48 file5.txt
drwxrwxr-x.  2 voxsteel voxsteel   23 Oct  8 17:48 testfolder
[voxsteel@centos8 packt]$ 
```

Figure 2.5 – cp and ls commands

- mv: You can use the mv command to move a file or directory from one location to another or even to rename a file. For example, you can rename a file from file1.txt to file2.txt by running the following command:

  ```
  mv file1.txt file2.txt
  ```

- rm: rm is used to remove files or directories, while the -r or -f parameter is used to recursively remove a directory (-r) or force remove a file or directory (-f). As always, use man to find out all the options possible.

- locate: The locate command is useful when you forget the location of a file. Using the -i argument helps to ignore case sensitivity. So, if you want to find a file named file1.txt, run the locate -i file1.txt command. This is the equivalent of search in Windows.

These are some basic commands that showed you how to list a file, check your working directory, create a directory, copy a file to another file, and so on. In the next section, we will use some more advanced commands.

Intermediate shell commands

In the previous section, we used some basic commands in order to get used to Terminal. In this section, we will get familiar with more advanced commands, as follows:

- echo: The echo command allows you to display content that can be added to either a new or an existing file or to replace the content.

- If you want to add content to an existing file, you can use echo "content to be appended" >>file1.txt. Or, you can use echo "this content will replace" > file1.txt to replace the content of a file.

  You can see the echo command in use here:

```
[voxsteel@centos8 packt]$ echo "Content to be added into file 1" >> file1.txt
[voxsteel@centos8 packt]$
```

Figure 2.6 – echo command

- cat: The cat command is normally used to read the content of a file, as illustrated here:

```
[voxsteel@centos8 packt]$ cat file1.txt
Content to overwrite the previous content
Content to be added into file 1
[voxsteel@centos8 packt]$
```

Figure 2.7 – cat command

You can use the cat command and append the output to a new file using >>. This is the same for any output—for example, you can use ls -la >> files-directories.txt to redirect the result of the ls -la command into a file.

- df: A great command for quickly viewing your filesystem and all mounted drives is the df command (which stands for *disk-free*). You can see the overall disk size, the amount of space used, the amount of space available, the utilization percentage, and the partition that the disk is mounted on. I advise using it along with the -h parameter to make the data legible by humans. The data that you see here was derived from the filesystem level or mount point:

```
[[voxsteel@centos8 packt]$ df -h
Filesystem              Size  Used Avail Use% Mounted on
devtmpfs                5.7G     0  5.7G   0% /dev
tmpfs                   5.8G     0  5.8G   0% /dev/shm
tmpfs                   5.8G  147M  5.6G   3% /run
tmpfs                   5.8G     0  5.8G   0% /sys/fs/cgroup
/dev/mapper/cl-root      70G  7.5G   63G  11% /
/dev/sdb2              1014M  450M  565M  45% /boot
/dev/mapper/cl-home     161G  2.7G  159G   2% /home
/dev/sdb1               599M  7.3M  592M   2% /boot/efi
/dev/sda1               469G   15G  430G   4% /data
tmpfs                   1.2G   40K  1.2G   1% /run/user/1000
[voxsteel@centos8 packt]$ 
```

Figure 2.8 – df command

- du: When used appropriately, the du command (which stands for disk usage) works great. When you need to know the size of a specific directory or subdirectory, this command excels. It only reports on the supplied stats at the time of execution and operates at the object level. For instance, you can use the du -sh /Documents command to find out how much disk space Linux's Documents folder consumes. This command works well when combined with the -sh flags to provide a summary of a given item in human-readable form (the directory and all subdirectories). You can see the du command in use here:

```
[[voxsteel@centos8 packt]$ du -sh /home/voxsteel/
442M    /home/voxsteel/
[voxsteel@centos8 packt]$ 
```

Figure 2.9 – du command

- uname: The uname command displays information regarding the operating system that your Linux distribution is currently using. The majority of the information on the system can be printed by using the uname -a command. This displays the kernel release date, the version, the processor type, and other related information. You can see the uname command in use here:

```
[[voxsteel@centos8 packt]$ uname -a
Linux centos8 4.18.0-348.7.1.el8_5.x86_64 #1 SMP Wed Dec 22 13:25:12 UTC 2021 x86_64 x86_64 x86_64 GNU/Linux
[voxsteel@centos8 packt]$ █
```

Figure 2.10 – uname command

- chmod: The system call and command used to modify the special mode flags and access permissions of filesystem objects are called chmod. These were first referred to collectively as its modes, and the name chmod was chosen as an acronym for change mode (more details about this in *Chapter 7, User and Group Management*).

Let's imagine you wish to change the permissions of a file called file1.txt so that the following is possible:

- The user can execute, read, and write it

- Those in your group can read it and use it

- Others may only read it

This command does the trick:

```
chmod u-rwx,g=rx,o=r file1.txt
```

The symbolic permissions notation is used in this example. u, g, and o stand for *user*, *group*, and *other*, respectively. The letters r, w, and x stand for *read*, *write*, and *execute*, respectively, while the equals sign (=) signifies *establish the permissions precisely like this*. There are no spaces between the various authorization classes; only commas are used to divide them.

Here is the equivalent command using octal permissions notation:

```
chmod 754 file1.txt
```

Here, the numbers 7, 5, and 4 stand for the user, group, and other permissions, respectively, in that sequence. Each digit is made up of the digits 4, 2, 1, and 0, which denote the following:

- 4 stands for *read*

- *Write* has the prefix 2

- 1 denotes *execute*

- 0 means *no authorization*

Therefore, 7 is made up of the permissions 4+2+1 (read, write, and execute), 5 (read, no write, and execute), and 4 (read, no write, and no execute).

- chown: To change the owner of system files and directories on Unix and Unix-like operating systems, use the chown command. This will change the ownership to the user (voxsteel) and group (voxsteel) for a specified filename or directory, as illustrated here:

```
chown voxsteel:voxsteel <filename or directory name>
```

Use `chgrp` if you're a non-privileged user and want to modify the group membership of a file you own.

- `chgrp`: A filesystem object's group can be changed to one to which they belong using the `chgrp` command, as illustrated in the following snippet. Three sets of access permissions are available for a filesystem object: one for owner, one for group, and one for others:

```
chgrp groupname <filename> (or directory name)
```

Searching the man pages with keywords can help you find a command even if you've forgotten its name. The `man -k keyword` is the syntax to use. Running this command on Terminal, for instance, will list Terminal-specific commands.

Summary

Redirection, Bash history, command aliases, command-line trickery, and more were some of the more sophisticated ideas related to shell commands that we covered in this chapter. Don't worry if you're having problems memorizing everything; it's normal to feel overwhelmed by the amount of information presented here at first. I've been using Linux professionally for over 15 years, and I still don't know everything there is to know about it.

In the next chapter, we will talk about filesystems, the differences between them, and the structure of the main system directories and what are they used for.

3

The Linux Filesystem

Files and filesystems are the topics of discussion in this chapter. The Unix ethos of *everything is a file* carries on in Linux, and while that's not true 100% of the time, most resources in Linux are actually files.

In this chapter, we will first define several relevant concepts. Then we will investigate Linux's implementation of the *everything is a file* concept. We will then cover the specialized filesystems the kernel makes use of to report data about running processes or attached hardware. We will then move on to normal files and filesystems, things you would typically associate with documents, data, and applications. Finally, we will explore standard filesystem operations and provide comparisons with other available alternatives. It is very important to know what limitations has each filesystem type in order for you to take the best decisions.

In this chapter, we will cover the following topics:

- The types of filesystems available and the differences between them
- The directory trees and standard directories
- How to mount/unmount filesystems

What is a filesystem?

A filesystem (or sometimes *file system*) governs the storage and retrieval of data in computers. Without a filesystem, all the data saved on a computer's hard drive would be jumbled together, making it impossible to find certain files. Instead, with a filesystem, data is easily isolated and identified by breaking it up into pieces and giving each piece a name. Each collection of data is referred to as a *file*, a name that originally came from information systems that use paper storage. A *filesystem* is a name for the organizational framework and logical principles used to handle the names and groups of bits of data.

In fact, there are a variety of filesystems available for Linux; if you're unsure which one to use, we'll present a thorough list of the filesystems that Linux supports.

What types of Linux filesystems are there?

Linux has a variety of filesystems to choose from, including the following:

- **ext**: The first filesystem constructed expressly for Linux was called *ext*, which is an acronym for *extended filesystem*, and was released in 1992. The primary objective of ext's developers was to increase the maximum size of editable files, which at the time was limited to 64 MB. The maximum file size grew to 2 GB as a result of the new metadata structure that was created. The maximum length of filenames was also increased at the same time to 255 bytes.

- **ext2**: This is also known as the *second expanded system*. ext2 was developed by Remy Card, just like the first one, and was intended to replace Linux's original extension system. It introduced innovations in fields such as storage capacity and overall performance. The maximum file size was notably increased to 2 TB, as opposed to the previous version's 2 GB. Still, filenames remained limited to 255 bytes long.

- **ext3**: ext3 is an upgraded version of ext2 and was first developed in 2001. The 2-TB maximum file size did not change, but ext3 was superior to ext2 in that it is a journaling filesystem. The 2-TB maximum file size does not change. This means that if the computer, hard drive(s), or both crash for any reason or encounter some type of power outage, the files can be repaired and restored upon rebooting using a separate log that contains the changes performed before the crash.

ext3 supports three levels of journaling:

- **Journal**: In the event of a power outage, the filesystem ensures effective filesystem recovery by writing both user data and metadata to the journal. Of the three ext3 journaling modes, this is the slowest. This journaling mode reduces the likelihood that any changes to any file in an ext3 filesystem will be lost.

- **Writeback**: When using the `data=writeback` mode, only metadata updates are logged in the journal. Data updates, on the other hand, are written directly to their respective locations on the disk without being logged in the journal first. This approach can provide better performance for write-intensive workloads because it reduces the overhead of journaling data.

  Pros:

  - Improved performance for write-heavy workloads since data is written directly to disk without the extra overhead of journaling

  - Lower write latency as data doesn't need to be written twice (to the journal and then to the filesystem)

  Cons:

  - Reduced data consistency during a system crash or power failure. Since data updates are not journaled, there's a possibility of data loss or inconsistency in case of a crash.

  - In scenarios where data integrity is critical (for example, databases), the `writeback` mode may not be the best choice because it prioritizes performance over data consistency.

- **Ordered**: This mode does not update related filesystem metadata; instead, it flushes changes from file data to disk before updating the associated filesystem metadata. This is ext3's default journaling mode. Only the files that were in the process of being written to the disk *disappear* in the event of a power outage. The architecture of the filesystem is undamaged.

- **ext4**: The *fourth extended system*, often known as ext4, was launched in 2008. This filesystem is commonly used as the default filesystem for the majority of Linux distributions since it overcomes a number of shortcomings that the third extended system had. Ext4 supports significantly larger filesystems and individual file sizes compared to Ext3. It can handle filesystems up to 1 exabyte (1 EB) and individual files up to 16 terabytes (16 TB). Additionally, a directory in ext4 can have up to 64,000 subdirectories (as opposed to 32,000 in ext3).

  Extents have replaced fixed blocks as the primary method of data allocation in ext4. An extent's beginning and end locations on the hard disk serve as a description of it. The number of pointers needed to represent the location of all the data in larger files can be greatly reduced because of the ability to express extremely long, physically contiguous files in a single inode pointer entry.

- **JFS**: JFS stands for *Journaled File System*. It is a 64-bit filesystem developed by IBM. In 1990, the first version of JFS (also known as *JFS1*) was introduced for use with IBM's AIX operating system.

  Data recovery after a power outage is simple and reliable. Additionally, compared to other filesystems, JFS requires less CPU power.

- **XFS**: SGI began working on XFS in the early 1990s with the intention of using it as the filesystem for their IRIX operating system. To tackle the most difficult computing challenges, XFS was designed as a high-performance 64-bit journaling filesystem. Large file manipulation and high-end hardware performance are strengths of XFS. In SUSE Linux Enterprise Server, XFS is the default filesystem for data partitions.

- **Btrfs**: Chris Mason created the **copy-on-write** (**COW**) filesystem known as Btrfs. It is based on Ohad Rodeh's COW-friendly B-trees. Btrfs is a logging-style filesystem that links the change after writing the block modifications in a new area as opposed to journaling them. New changes are not committed until the last write.

- **Swap**: When the amount of memory available to the computer begins to run low, the system will use a file known as a swap file to generate temporary storage space on a solid-state drive or hard disk. The file replaces a section of memory in the RAM storage of a paused program with a new part, making memory available for use by other processes.

  The computer is able to utilize more RAM than is physically installed by using a swap file. In other words, it is capable of running more programs than it would be able to execute if it relied solely on the limited resources provided by the RAM that was physically installed.

  Because swap files are not kept in the computer's actual RAM, we can consider them to be a form of virtual memory. When a computer uses a swap file, its operating system essentially tricks itself into thinking that it has more RAM than it actually does.

Linux is compatible with a wide variety of filesystems, including the FAT and NTFS filesystems that are standard to other operating systems such as Windows. It's possible that embedded developers will support those, although in most cases, a Linux filesystem such as ext4, XFS, or Btrfs will be used for storage partitions. A better understanding of the benefits of the available alternatives will help you to choose the appropriate filesystem for a certain application.

High scalability

By leveraging allocation groups, XFS provides excellent scalability.

The block device that supports the XFS filesystem is split into eight or more linear regions that are all the same size at the moment the filesystem is created. They are referred to as allocation groups. Each allocation group controls its own free disk space and inodes. The kernel can address multiple allocation groups at once since they are relatively independent of one another. The high scalability of XFS is made possible by this feature. These autonomous allocation groups naturally meet the requirements of multiprocessor systems.

High performance

XFS provides high performance by effectively managing disk space.

Within the allocation groups, B+ trees manage free space and inodes. The effectiveness and scalability of XFS are considerably enhanced by the usage of B+ trees. XFS manages allocation in a delayed manner by dividing the allocation procedure into two steps. Pending transactions are kept in the RAM and the necessary amount of space is set aside. The precise location (in filesystem blocks) of the data's storage is still left up to XFS. This choice is postponed until the very last second. If it is outdated when XFS selects where to save it, certain short-lived temporary data may never reach the disk. XFS improves write performance and lessens filesystem fragmentation in this way. Data loss after a crash during a write operation is likely to be more severe in a delayed-allocation filesystem than in other filesystems.

What filesystem does my system use?

If you aren't sure which filesystem your distribution ships with, or if you just want to know which one you have installed, you can use some clever commands at the Terminal to find out.

There are other ways to accomplish this, but we'll demonstrate the simplest one here using the df -T command.

```
[/ $df -T
Filesystem              Type   1K-blocks       Used Available Use% Mounted on
tmpfs                   tmpfs    2437252       2576   2434676   1% /run
/dev/nvme0n1p2          ext4   244506940  21683604 210330280  10% /
tmpfs                   tmpfs   12186244          0  12186244   0% /dev/shm
tmpfs                   tmpfs       5120          4      5116   1% /run/lock
tmpfs                   tmpfs   12186244          0  12186244   0% /run/qemu
/dev/nvme0n1p1          vfat      523248       5380    517868   2% /boot/efi
/dev/mapper/cs-home ext4   409357880  2700504 385789692   1% /data
tmpfs                   tmpfs    2437248       4712   2432536   1% /run/user/1000
/ $
```

Figure 3.1 – A command used to determine what type of filesystem is in use

In the second column, labeled Type, you can see the descriptions of the filesystem formats. At this point, you should be able to tell which filesystem is mounted on your Linux installation.

FUSE filesystem

As a user, you shouldn't have to worry too much about the underlying implementation when interacting with files and directories in user space. It is common practice for processes to make use of system calls to the kernel in order to read or write to a mounted filesystem. However, you do have access to data from the filesystem that doesn't seem to belong in the user's domain. The stat() system call in particular returns inode numbers and link counts.

Do you have to worry about inode numbers, link counts, and other implementation details even when you're not maintaining a filesystem? No (in most cases). This information is made available to user-mode programs for the primary purpose of maintaining backward compatibility. In addition, these filesystem internals aren't present in every Linux filesystem because they're not standardized. The VFS interface layer is responsible for ensuring that system calls always return inode numbers and link counts; however, the values of these numbers may or may not indicate anything.

On non-traditional filesystems, it's possible that you won't be able to carry out operations that are typical of the Unix filesystem. For instance, you cannot use the ln command to create a hard link on a mounted VFAT filesystem because that filesystem's directory entry structure does not support the concept of hard links. Because of the high level of abstraction provided by the system calls available in user space, working with files on Linux systems does not require any prior knowledge of the underlying implementation. Additionally, the format of filenames is flexible, and the use of mixed-case filenames is supported; both of these features make it simple to support other hierarchical-style filesystems.

> **Note**
>
> Keep in mind that the support for a particular filesystem does not necessarily need to be included in the kernel. To give one illustration, the role of the kernel in user-space filesystems is limited to that of a conduit for system calls.

The directory tree and standard directories

To see the main structure of the root folder, just use the following command: `tree -L 1`.

Figure 3.2 – The command to see a directory structure tree

To better grasp how the Linux filesystem functions in general, let's examine what each folder's purpose is with reference to the Linux filesystem diagram shown in *Figure 3.2*. Not all of the folders mentioned here nor in the preceding examples will be found in every Linux distribution, but the vast majority of them will:

- `/bin`: The majority of your binary files are kept in this location, which is pronounced *bin*, and is often used by Linux Terminal commands and essential utilities such as `cd` (change directory), `pwd` (print working directory), `mv` (move), and others.

- /boot: All of the boot files for Linux can be found in this folder. The majority of people, including myself, save this directory on a separate partition of their hard drive, especially if they plan to use dual-booting. Remember that even if /boot is physically located on a different partition, Linux still believes it to be at /boot.

- /dev: Your physical devices, such as hard drives, USB drives, and optical media, are mounted here. Additionally, your drive may have different partitions, in which case you'll see /dev/sda1, /dev/sda2, and so forth.

- /etc: This directory stores configuration files. Users can keep configuration files in their own /home folder, which affects only the given user, whereas configurations placed in /etc usually affect all users on the system.

- /home: Because this directory contains all of your personal information, you'll spend most of your time here. The /home/username directory contains the Desktop, Documents, Downloads, Photos, and Videos directories.

- /lib: Here is where you'll find all the library buildings. There are always extra libraries that start with lib-something that get downloaded when you install a Linux distribution. The operation of your Linux program depends on these files.

- /media: This is where external devices such as USB drives and CD-ROMs are mounted. This varies between Linux distributions.

- /mnt: This directory basically serves as a mounting point for other folders or drives. This can be used for anything, although it is typically used for network locations.

- /opt: This directory contains supplementary software for your computer that is not already managed by the package management tool that comes with your distribution.

- /proc: The *processes* folder contains a variety of files holding system data (remember, everything is a file). In essence, it gives the Linux kernel—the heart of the operating system—a mechanism to communicate with the numerous processes that are active within the Linux environment.

- /root: This is the equivalent of the /home folder for the root user, commonly known as the superuser. You should only touch anything in this directory if you are really sure you know what you're doing.

- /sbin: This is comparable to the /bin directory, with the exception that it contains instructions that can only be executed by the *root* user, sometimes known as the *superuser*.

- /tmp: Temporary files are kept here and are often erased when the computer shuts down, so you don't have to manually remove them as you would in Windows.

- /usr: This directory contains files and utilities that are shared between users.

- /var: The files used by the system to store information as it runs are often located in the /var subfolder of the root directory in Linux and other Unix-like operating systems.

We've now covered the root directory, but many of the subdirectories also lead to additional files and folders. You can get a general concept of what the basic filesystem tree looks like from the following diagram:

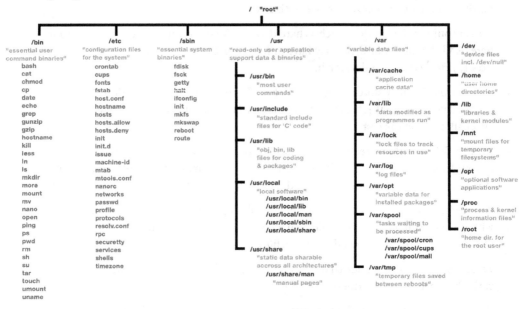

Figure 3.3 – Basic filesystem structure (Source: https://en.wikipedia.org/wiki/Unix_filesystem)

Having an understanding of the root directory structure will make your life much easier in the Linux world.

Links (hard and symbolic)

There are two alternative ways to refer to a file on the hard drive: hard links and symbolic links. The filesystem, which organizes which file belongs where and how, includes several approaches such as symbolic link and hard link. A hard link basically refers to the inode of a file and is a synchronized carbon copy of that file. On the other hand, symbolic links point directly to the file, which in turn points to the inode, a shortcut. We need to next discuss inodes in order to comprehend how symbolic and hard links function.

What is an inode?

A Unix-style filesystem uses a data structure called an inode to describe filesystem objects such as files and directories. The properties and disk block locations of an object's data are stored in each inode. Attributes of filesystem objects can include metadata, owner information, and permission information.

Inodes are essentially a whole address's numerical equivalent. The operating system can obtain details about a file, including permission privileges and the precise location of the data on the hard drive, using an inode.

What is a hard link?

On a computer system, a hard link is a special kind of link that points directly to a specific file by its name. A hard link will continue to point to the original file even if the file's name is changed, unlike a soft link.

When comparing the two methods of linking a directory entry or file to the same memory region, hard links are more reliable. As opposed to symbolic links, hard links prevent files from being deleted or moved. The *alias effect*, in which a file has numerous identifiers, can occur when multiple hard links point to the same file. Experts agree that all links, whether hard or soft, are pointers; nonetheless, hard links are distinguished as being more permanent than soft connections.

What are symbolic links?

Symbolic links are essentially shortcuts that refer to a file rather than the inode value of the file they point to. This method can be applied to directories and can be used to make references to data that is located on a variety of hard discs and volumes. A symbolic link will be broken or leave a dangling link if the original file is moved to a different folder. This is due to the fact that symbolic links refer to the original file and not the inode value of the file.

Because a symbolic link points to the original file, any changes you make to the symbolic link should result in corresponding changes being made to the actual file.

Mounting and unmounting filesystems

In order for the computer to access files, the filesystem must be mounted. The mount command will show you what is mounted (usable) on your system at the moment.

I created my own /data folder and mounted a new HDD into it:

```
[/ $mount | grep data
/dev/mapper/cs-home on /data type ext4 (rw,relatime,errors=remount-ro)
```

Figure 3.4 – A command showing what filesystem is mounted on /data

To mount your filesystem with a command, just run the following:

```
mount -t ext4 /dev/mapper/cs-home /data
```

In order to have it automatically mounted on reboot, you have to define this entry in `/etc/fstab`.

If you want to mount a CD-ROM, just run the following command:

```
mount -t iso9660 /dev/cdrom /mnt/cdrom
```

For more detailed information, consult the `mount` man page or run `mount` with the `-h` flag to get assistance.

The `cd` command can be used to traverse the newly accessible filesystem through the mount point you just created after mounting.

How to unmount the filesystem

Using the `umount` command and specifying the mount point or device, you can unmount (detach) the filesystem from your computer.

The following command, for instance, would unmount a CD-ROM:

```
umount /dev/cdrom
```

Pseudo-filesystems

A process information pseudo-filesystem is another name for the `proc` filesystem. It contains runtime system information rather than *actual* files (for example, system memory, devices mounted, hardware configuration, and so on). It can therefore be viewed as the kernel's command and information hub. This directory (`/proc`) is accessed by many system utilities. The `lsmod` command lists the modules loaded by the kernel, and the `lspci` command displays the devices attached to the PCI bus. Both of these commands are functionally equivalent to `cat /proc/modules` and `cat /proc/pci`, respectively. Common examples of pseudo filesystems in Unix-like operating systems (for example, Linux) include the following:

- Processes, the most prominent use
- Kernel information and parameters
- System metrics, such as CPU usage

Processes

All the information about each running process can be found in the `/proc/pid` file. Here's an illustration of a few PIDs in action right now:

```
[~ $ls -l /proc |head -10
total 0
dr-xr-xr-x  9 root           root              0 Sep  7 10:58 1
dr-xr-xr-x  9 root           root              0 Sep  7 10:58 10
dr-xr-xr-x  9 root           root              0 Sep  7 10:58 1018
dr-xr-xr-x  9 root           root              0 Sep  7 10:58 1031
dr-xr-xr-x  9 root           root              0 Sep  7 10:58 1059
dr-xr-xr-x  9 root           root              0 Sep  7 10:58 1061
dr-xr-xr-x  9 root           root              0 Sep  7 10:58 1064
dr-xr-xr-x  9 colord         colord            0 Sep  7 10:58 1065
dr-xr-xr-x  9 root           root              0 Sep  7 10:58 11
~ $
```

Figure 3.5 – A command to see running processes

Let's take, for example, PID 1031 and see what is inside:

```
~ $sudo ls -l /proc/1031
total 0
-r--r--r--  1 root root 0 Sep  7 11:04 arch_status
dr-xr-xr-x  2 root root 0 Sep  7 10:58 attr
-rw-r--r--  1 root root 0 Sep  7 11:04 autogroup
-r--------  1 root root 0 Sep  7 11:04 auxv
-r--r--r--  1 root root 0 Sep  7 10:58 cgroup
--w-------  1 root root 0 Sep  7 11:04 clear_refs
-r--r--r--  1 root root 0 Sep  7 10:58 cmdline
-rw-r--r--  1 root root 0 Sep  7 10:58 comm
-rw-r--r--  1 root root 0 Sep  7 11:04 coredump_filter
-r--r--r--  1 root root 0 Sep  7 11:04 cpu_resctrl_groups
-r--r--r--  1 root root 0 Sep  7 11:04 cpuset
lrwxrwxrwx  1 root root 0 Sep  7 11:00 cwd -> /
-r--------  1 root root 0 Sep  7 11:04 environ
lrwxrwxrwx  1 root root 0 Sep  7 10:58 exe -> /usr/sbin/sshd
dr-x------  2 root root 0 Sep  7 10:58 fd
dr-xr-xr-x  2 root root 0 Sep  7 11:04 fdinfo
-rw-r--r--  1 root root 0 Sep  7 11:04 gid_map
-r--------  1 root root 0 Sep  7 11:04 io
-r--r--r--  1 root root 0 Sep  7 11:04 limits
-rw-r--r--  1 root root 0 Sep  7 10:58 loginuid
dr-x------  2 root root 0 Sep  7 11:04 map_files
-r--r--r--  1 root root 0 Sep  7 11:00 maps
-rw-------  1 root root 0 Sep  7 11:04 mem
-r--r--r--  1 root root 0 Sep  7 11:04 mountinfo
-r--r--r--  1 root root 0 Sep  7 11:04 mounts
-r--------  1 root root 0 Sep  7 11:04 mountstats
dr-xr-xr-x 60 root root 0 Sep  7 11:04 net
dr-x--x--x  2 root root 0 Sep  7 11:04 ns
-r--r--r--  1 root root 0 Sep  7 11:04 numa_maps
-rw-r--r--  1 root root 0 Sep  7 11:04 oom_adj
-r--r--r--  1 root root 0 Sep  7 11:04 oom_score
-rw-r--r--  1 root root 0 Sep  7 10:58 oom_score_adj
```

Figure 3.6 – A command to see what is inside of the process with PID 1031

Finally, a synthetic filesystem is a filesystem that provides a tree-like interface to non-file objects, making them look like regular files in a disk-based or long-term storage filesystem. This type of filesystem is also known as a faux filesystem.

Kernel and system information

Numerous folders under /proc contain a wealth of knowledge about the kernel and the operating system. There are too many of them to include here, but we will cover a few along with a brief description of what they contain:

- /proc/cpuinfo: Information about the CPU
- /proc/meminfo: Information about the physical memory
- /proc/vmstats: Information about the virtual memory
- /proc/mounts: Information about the mounts
- /proc/filesystems: Information about filesystems that have been compiled into the kernel and whose kernel modules are currently loaded
- /proc/uptime: This shows the current system uptime
- /proc/cmdline: The kernel command line

CPU usage

When evaluating a system's overall performance, it is crucial to have a firm grasp of how much CPU time is being used. Knowing how to monitor CPU utilization in Linux via the command line is essential knowledge for everyone working with Linux, from enthusiasts to system administrators.

One of the most common commands used for this purpose is top:

```
top - 11:53:15 up 33 days, 21:46,  2 users,  load average: 1.25, 1.16, 1.16
Tasks: 358 total,   1 running, 357 sleeping,   0 stopped,   0 zombie
%Cpu(s):  3.2 us,  1.0 sy,  0.0 ni, 94.7 id,  0.1 wa,  0.7 hi,  0.2 si,  0.0 st
MiB Mem :  11681.1 total,   1112.4 free,   8409.2 used,   2159.4 buff/cache
MiB Swap:   6004.0 total,   3652.7 free,   2351.2 used.   2779.2 avail Mem

    PID USER      PR  NI    VIRT    RES    SHR S  %CPU  %MEM     TIME+ COMMAND
   5503 pipewire  20   0   42948  19784  14480 S   5.3   0.2 540:27.19 postgres
   5123 polkitd   20   0 1628572 727212  15816 S   2.3   6.1 2534:53 bundle
1171268 polkitd   20   0 1157116 661056  13148 S   1.7   5.5   0:10.29 bundle
   5044 geoclue   20   0   70904  20848   3616 S   1.3   0.2 757:27.73 redis-server
   5159 polkitd   20   0  199464  41292   9864 S   1.0   0.3 382:26.49 gitlab-exporter
   1050 root      20   0  297200   7564   6728 D   0.7   0.1 141:13.36 iio-sensor-prox
   1048 dbus      20   0   74156   6896   4396 S   0.3   0.1  57:27.48 dbus-daemon
   1272 root      20   0  409636  16308  13032 S   0.3   0.1  37:16.26 NetworkManager
   1303 root      20   0 1510500  30788   4476 S   0.3   0.3  28:43.65 containerd
   1596 haproxy   20   0  158864   1388    452 S   0.3   0.0   9:17.53 haproxy
   3431 root      20   0  712136   9692      0 S   0.3   0.1  12:11.50 containerd-shim
   5185 saslauth  20   0 3441836 237624  26664 S   0.3   2.0 347:13.48 prometheus
   5224 saslauth  20   0 1524852  44240  10680 S   0.3   0.4  13:35.69 grafana-server
```

Figure 3.7 – Output of top command

The system's response should be a comprehensive rundown of all active tasks where users, processes, CPU usage, and memory consumption can all be viewed.

Summary

In this chapter, we explored files and filesystems in Linux. Filesystems are a powerful and versatile approach to organize access to information in a hierarchical fashion. In Linux, filesystems are the focus of numerous technologies and ongoing efforts. Some are open source, but there is also a spectrum of commercial options.

In the next chapter, we will talk about processes and process control.

Processes and Process Control

The main role of an operating system is to allow applications to run on a computer and use its resources. A lot of the time, the role of the systems administrator boils down to making sure that the right processes are running and diagnosing if they are not running. Thus, it is important to understand how the operating system starts and handles processes and how to start, stop, and monitor them.

In this chapter, we will cover the following topics:

- Executables versus processes
- Process termination, exit codes, and signals
- The process tree
- Process search and monitoring

Executables versus processes

Programs are distributed as *executable files*. In many historical operating systems, programs would be loaded from files directly into memory byte by byte. That approach was certainly simple to implement, but it has many limitations (most notably, the requirement to have a fixed memory layout and the inability to store any metadata), so later systems invented special formats for executable files.

For example, if we inspect the **Bourne Again Shell** (**Bash**) executable with the file command, we'll see something like this:

```
$ file /bin/bash
/bin/bash: ELF 64-bit LSB pie executable, x86-64, version 1 (SYSV),
dynamically linked, interpreter /lib64/ld-linux-x86-64.so.2,
BuildID[sha1]=9c4cb71fe5926100833643a8dd221ffb879477a5, for GNU/Linux
3.2.0, stripped
```

If you use a Linux distribution other than Debian or Red Hat derivatives (which are the main focus of this book) and the preceding command fails for you, you can find the location of the bash executable with which bash, or choose a different executable, such as cat or ls.

ELF stands for **Executable and Linkable Format**. It's the default executable file format on Linux and many other Unix-like operating systems (**LSB** means **Linux Standard Base**). An ELF file stores executable code of programs – machine instructions that are loaded into memory to be executed by the CPU. However, it can also store debug information, such as associations between machine instructions and lines in the program source code they were compiled from. ELF files can also be *linked* with other ELF files, known as *shared libraries* – files that contain executable code but aren't meant to run as programs and only serve as collections of reusable functions.

You can see the library linkage information with the `ldd` command:

```
$ ldd /bin/bash
    linux-vdso.so.1 (0x00007ffc06ddb000)
    libtinfo.so.6 => /lib64/libtinfo.so.6 (0x00007f30b7293000)
    libc.so.6 => /lib64/libc.so.6 (0x00007f30b7000000)
    /lib64/ld-linux-x86-64.so.2 (0x00007f30b743e000)
```

If you run `file` on `libc.so.6`, the standard library for the C programming language, you will see that it's also an ELF file:

```
$ file /lib64/libc.so.6
/lib64/libc.so.6: ELF 64-bit LSB shared object, x86-64, version 1
(GNU/Linux), dynamically linked, interpreter /lib64/ld-linux-x86-64.
so.2, BuildID[sha1]=6f5ce514a9e7f51e0247a527c3a41ed981c04458, for GNU/
Linux 3.2.0, not stripped
```

Finally, ELF stores metadata such as the target operating system and CPU architecture. The file command doesn't guess that files from the examples earlier are for Linux on x86-64, and simply gets them from the ELF file header.

In the output of `file /bin/bash`, you might have noticed an unusual field – `interpreter /lib64/ld-linux-x86-64.so.2`. Bash is written in C, which is a compiled language, and shouldn't need any interpreter. Indeed, that executable contains machine code, and the Linux kernel knows how to load ELF files; if it didn't, it couldn't load that `ld-linux` interpreter, resulting in a chicken-and-egg problem.

The role of `ld-linux.so` is not to interpret the executable itself but, instead, to correctly resolve references to functions that come from shared libraries. If you run `file` on it, you will see that it's `static-pie` linked rather than `dynamically` `linked`, unlike `/bin/bash` — `static-pie` means `static position-independent executable`:

```
$ file /lib64/ld-linux-x86-64.so.2
/lib64/ld-linux-x86-64.so.2: ELF 64-bit LSB shared
object, x86-64, version 1 (GNU/Linux), static-pie linked,
BuildID[sha1]=8fa0bd5df5fa6fff60bb4cbdd753621d00f94dfc, with debug_
info, not stripped
```

The kernel knows nothing about programs' library function dependencies and can only load statically linked ELF executables directly. To load dynamically linked executables, it relies on the `ld-linux.so` helper but reuses a general interpreter association mechanism for it, instead of inventing something custom just for that case.

Programs written in interpreted languages such as Python, Ruby, or shell require an actual interpreter to be loaded first. This is specified using a *shebang line* that starts with `#!`.

You can try it yourself by creating a simple shell script:

```
$ echo '#!/bin/bash' >> hello.sh
$ echo 'echo "hello world"' >> hello.sh
$ chmod +x ./hello.sh
$ ./hello.sh
hello world
```

If a file has an executable bit (+x) on it and starts with a shebang line, the kernel will first load its interpreter (in this case, `/bin/bash`) and then give it the executable as an argument.

Once an executable file is loaded, directly by the kernel itself or with help from an interpreter, it becomes a running *process*.

Process termination and exit codes

All processes have to eventually terminate, and there are many situations when process execution cannot continue, either due to errors in its own program logic or problems with the environment (such as missing files, for example). The user may also need to terminate processes by hand, either to make changes to the system, or to prevent a misbehaving process from taking up resources or interfering with the system's functioning.

In this section, we will learn how to examine the exit code of a terminated process to guess the possible reasons for its termination, and how to communicate with processes and force their termination.

Exit codes

Most processes are short-lived – they do their job and terminate by themselves. Every process terminates with an *exit code* – a numeric value that indicates whether it exited normally or terminated due to an error. By convention, a zero exit code means success, and any non-zero exit code indicates an error. There are no standard meanings for non-zero exit codes – exact meanings vary between programs and operating systems, and many programs simply exit with 1 if they encounter an error, no matter what that error is.

In bash, you can find the exit code of the last command in a special variable named $?. There's a pair of Unix commands whose sole purpose is to exit with success and error codes respectively, `true` and `false`:

```
$ true
$ echo $?
0
$ false
$ echo $?
1
```

Most of the time, programs set their exit codes themselves. For example, in a shell script, you can use `exit 1` to signal an error. In C, you can use `return 1` in your `main()` function to the same effect. For programs that can be executed non-interactively from scripts, it's critically important to exit with a non-zero code on errors; otherwise, script authors will have no way to know whether their script steps succeeded or failed.

All standard commands do this. For example, let's try to create a file in `/etc/` from a normal user and see what it leaves in the $? variable:

```
$ touch /etc/my_file
touch: cannot touch '/etc/my_file': Permission denied
$ echo $?
1
```

The simplest use case for exit codes is chaining commands with the || and && operators. They can be called *on error* and *on success* – in `cmd1 || cmd2`, the shell will execute `cmd2` if `cmd1` fails (that is, exits with a non-zero code). In `cmd1 && cmd2`, it's the other way around – `cmd2` is only executed if `cmd1` succeeds (exits with zero code):

```
$ touch /etc/my_file || echo "Fail!"
touch: cannot touch '/etc/my_file': Permission denied
Fail!
$ touch /tmp/my_file && echo "Success!"
Success!
```

On errors such as the file permission error in our example, the kernel simply does not do what the program asks it to do and, instead, allows the program to continue running as usual. The reasoning is that such errors often occur due to incorrect user input while the program logic is correct, so the program needs to be able to handle them and notify the user. However, in other cases, the kernel will interrupt a process by generating a *signal*.

Signals

A signal is a special condition that may occur during process execution. There are many signals defined in the POSIX standard. Some are associated with specific program logic errors, such as SIGILL — *illegal instruction* (caused, for example, by attempts to divide by zero) — or SIGSEV — *segmentation violation* (caused by trying to read or modify memory that wasn't allocated to the process). Other signals are generated on external conditions to force a process to handle them, such as SIGPIPE, which is generated when a network socket or a local pipe is closed by the other end. These signals are only of interest to software developers, but some are designed as process control tools for administrators, such as SIGINT (which interrupts a process), SIGTERM (which asks the process to clean up its state and terminate), and SIGKILL (which tells the kernel to forcibly terminate a process). It's said that a signal is *sent to a process*. That terminology is a good abstraction for users, but in reality, it's the kernel that has execution control when a signal is generated, not the process. Programmers may anticipate certain signals and register *signal handlers* for them. For example, many interpreters for high-level programming languages handle the SIGILL signal and convert it into exceptions such as ZeroDivisionError in Python. However, if a programmer forgot or chose not to register a handler for SIGILL and the program attempts to divide by zero, the kernel will terminate the process.

If you have **GNU Compiler Collection (GCC)** installed, you can see it for yourself with a simple C program:

```
$ echo "void main() { int x = 0 / 0; }" >> sigill.c
$ gcc -o sigill ./sigill.c
./sigill.c: In function 'main':
./sigill.c:1:25: warning: division by zero [-Wdiv-by-zero]
    1 | void main() { int x = 0 / 0; }
      |                          ^
$ ./sigill
Floating point exception (core dumped)
$ echo $?
136
```

GCC helpfully warns you that your program is incorrect, but if you still try to run it, the kernel forcibly terminates it and sets the error code to a non-zero value.

Signals such as SIGILL or SIGPIPE occur regardless of the user's wishes, but there's also a class of signals that is meant to be initiated by users (or by processes on their behalf).

The kill command

The command for sending signals to processes is called **kill**. That name is also somewhat misleading; most often, it's indeed used to forcibly terminate processes, but it can also send other signals as well.

To illustrate its usage, let's learn how to send a process to the background. In bash, you can do that by appending an ampersand to a command. Using the jobs command, you can see a list of background processes, and by using fg <job number>, you can bring a job with a certain number to the foreground. Here's how you can send a cat process to the background and then bring it back:

```
$ cat &
[1] 22501

[1]+  Stopped                  cat
$ jobs
[1]+  Stopped                  cat
$ fg 1
cat
hello world
hello world
^C
$
```

When you press *Ctrl* + *C* to terminate a process in the shell, you actually ask your shell to send it a SIGINT signal – a signal to interrupt execution. If a process is in the background, we cannot use *Ctrl* + *C* to interrupt it. However, with kill, we can send it manually:

```
$ cat &
[1] 22739

[1]+  Stopped                  cat
$ kill 22739
$ jobs
[1]+  Stopped                  cat
$ fg 1
cat
Terminated
```

Here's what happened – when we ran cat &, the shell told us its *background job number* (1) and also its **process identifier** (often abbreviated as **PID**); in this case, the number was 22739, but it can be any number. We then used kill 22739 to send the process a signal, and indeed, when we tried to bring it to the foreground, the shell told us that it terminated while it was in the background.

By default, the `kill` command sends a `SIGTERM` signal. Both `SIGINT` and `SIGTERM` can be caught or ignored by a process. By sending them to a process, you ask it to terminate; a well-behaved process should comply and may use it as a chance to finalize its current task before terminating – for example, to finish writing data to a file. That means that neither the *Ctrl* + *C* key combination in the shell nor the `kill` command with default options is suitable for forcibly terminating a misbehaving process.

To force a process to quit, you should use `kill -SIGKILL <PID>` instead, or its numeric equivalent, `kill -9 <PID>`. However, it should be your last resort since the kernel will simply end the process immediately, and that may leave its files in an inconsistent state.

If you run `kill -l`, you will see a long list of available signals:

```
$ kill -l
 1) SIGHUP       2) SIGINT       3) SIGQUIT      4) SIGILL       5) SIGTRAP
 6) SIGABRT      7) SIGBUS       8) SIGFPE       9) SIGKILL     10) SIGUSR1
11) SIGSEGV     12) SIGUSR2     13) SIGPIPE     14) SIGALRM     15)
SIGTERM
16) SIGSTKFLT   17) SIGCHLD     18) SIGCONT     19) SIGSTOP     20)
SIGTSTP
21) SIGTTIN     22) SIGTTOU     23) SIGURG      24) SIGXCPU     25) SIGXFSZ
26) SIGVTALRM   27) SIGPROF     28) SIGWINCH    29) SIGIO       30)
SIGPWR
31) SIGSYS      34) SIGRTMIN    35) SIGRTMIN+1  36) SIGRTMIN+2  37)
SIGRTMIN+3
38) SIGRTMIN+4  39) SIGRTMIN+5  40) SIGRTMIN+6  41)
SIGRTMIN+7      42) SIGRTMIN+8
43) SIGRTMIN+9  44) SIGRTMIN+10  45) SIGRTMIN+11  46)
SIGRTMIN+12     47) SIGRTMIN+13
48) SIGRTMIN+14  49) SIGRTMIN+15  50) SIGRTMAX-14  51)
SIGRTMAX-13     52) SIGRTMAX-12
53) SIGRTMAX-11  54) SIGRTMAX-10  55) SIGRTMAX-9   56)
SIGRTMAX-8      57) SIGRTMAX-7
58) SIGRTMAX-6  59) SIGRTMAX-5   60) SIGRTMAX-4   61)
SIGRTMAX-3      62) SIGRTMAX-2
63) SIGRTMAX-1  64) SIGRTMAX
```

Some of those signals have no predefined meanings and, instead, are specific to applications. For example, in `SIGUSR1` and `SIGUSR2`, USR stands for *user-defined*. Most processes ignore them, but some use them to allow system administrators to force a config reload or another operation.

Now, we know how to examine the error codes of terminated processes and get an idea of whether they exited normally or failed. We also learned how the operating system kernel communicates with processes using signals, and how we can use the `kill` command to ask the kernel to either tell a process to exit or terminate a misbehaving process forcibly. Now, let's learn how to explore the running processes and relationships between them.

The process tree

We've seen that the shell knows the PIDs of the commands you run and can send them signals to terminate when you press *Ctrl + C*. That implies that it has certain control over the processes you ask it to launch. Indeed, everything you launch from your shell becomes a *child process* of that shell process.

The shell itself is a child process — either of your terminal emulator if you are on a Linux desktop, or of the OpenSSH daemon if you connect remotely over SSH. However, is there a parent of all processes, or can there be multiple processes without parents?

In fact, there is a parent of all processes, and all running process relationships form a tree with a single root (PID = 1). For historical reasons, the parent of all processes is often called the *init process*. For a long time in general-purpose Linux distributions, that process was System V init, hence the term.

The PID=1 process can be anything. When you boot a Linux system, you can tell it which executable to load as PID=1. For example, one way to boot a system in rescue mode is to append `init=/bin/bash` to the GRUB command line (but you are better off using a built-in rescue option in your distro's boot menu item because it may pass additional useful parameters). That will make your kernel drop into a single-user shell session instead of initiating its normal boot process. Some embedded systems that use Linux just as a hardware abstraction layer may start custom processes instead. However, normally, the process with PID=1 serves as a service manager.

The System V init served as the de facto standard service manager for over two decades, but most modern distributions use systemd instead, while some opt for other alternatives to the old System V init, such as OpenRC.

The `init` process is the only process that is launched directly by the kernel. Everything else is launched by the `init` process instead: login managers, the SSH daemon, web servers, database systems – everything you can think of. You can view the full process tree with the `pstree` command. Here's a tree from a small web server:

```
$ pstree
systemd─┬─NetworkManager───2*[{NetworkManager}]
        ├─2*[agetty]
        ├─auditd───{auditd}
        ├─chronyd
        ├─crond
        ├─dbus-broker-lau───dbus-broker
        ├─do-agent───5*[{do-agent}]
        ├─droplet-agent───8*[{droplet-agent}]
        ├─nginx───nginx
        ├─sshd───sshd───sshd───bash───pstree
        ├─systemd───(sd-pam)
        ├─systemd-homed
        ├─systemd-hostnam
```

```
        ├─systemd-journal
        ├─systemd-logind
        ├─systemd-oomd
        ├─systemd-resolve
        ├─systemd-udevd
        └─systemd-userdbd───3*[systemd-userwor]
```

Here, you can see that `pstree` was a child process of a bash session, which in turn was a child of `sshd` (an OpenSSH process), which was itself a child of `systemd` – the root of the process tree.

However, most of the time, you will be interested in finding specific processes and their resource usage.

Process search and monitoring

The `pstree` command is a great way to visualize all running processes and relationships between them, but in practice, most of the time administrators look for specific processes or need to learn about their resource usage, rather than their mere existence. Let's learn about the tools for those tasks – the `ps` command to search processes, the `top` command to monitor resource usage in real time, and the underlying kernel interface that those tools use.

The ps command

PS is an abbreviation for **process selection** or **process snapshot**. It's a utility that allows you to retrieve and filter information about running processes.

Running `ps` without any arguments will get you a very limited selection – only processes that run from your user and that are attached to a *terminal* (that is, aren't processes that run with all input and output closed and only communicate with other processes through network or local sockets — usually daemons, but GUI programs may behave the same way).

Somewhat confusingly, `ps` itself always shows up in such lists because, when it gathers that information, it's indeed a running process:

```
$ ps
    PID TTY          TIME CMD
 771681 pts/0    00:00:00 bash
 771705 pts/0    00:00:00 ps
```

The `PID` field is, of course, a process identifier – a unique number that the kernel assigns to every process when that process is launched. If present, the `TTY` field is a terminal – it can be a real serial console (usually `ttyS*` or `ttyUSB*`), a virtual console on a physical display (`tty*`), or a purely virtual pseudo-terminal associated with an SSH connection or a terminal emulator (`pts/*`).

The CMD field shows the command that was used to launch a process with its arguments, if any were used.

The `ps` command has a large number of options. Two options you should learn about right away are a and x — options that remove *owned by me* and *have a terminal* restrictions.

A common command to view every process on the system is `ps  ax`. Let's try to run it:

```
$ ps ax
    PID TTY       STAT     TIME COMMAND
    1 ?           Ss     1:13 /usr/lib/systemd/systemd --switched-root
--system --deserialize 30
    2 ?           S        0:01 [kthreadd]
    3 ?           I<       0:00 [rcu_gp]
    4 ?           I<       0:00 [rcu_par_gp]
                  ...
    509 ?         Ss       7:33 /usr/lib/systemd/systemd-journald
    529 ?         S        0:00 [jbd2/vda2-8]
    530 ?         I<       0:00 [ext4-rsv-conver]
    559 ?         Ss     226:16 /usr/lib/systemd/systemd-oomd
    561 ?         Ss       0:49 /usr/lib/systemd/systemd-userdbd
    562 ?         S<sl     1:17 /sbin/auditd
    589 ?         Ss       0:10 /usr/lib/systemd/systemd-homed
    590 ?         Ss       0:02 /usr/lib/systemd/systemd-logind
```

The `STAT` field tells us the process state. The S state means a process is in the interruptible sleep state – it waits for external events. Processes that currently do something can be seen in the R state – running. The I state is special; it is for idle kernel threads. The most concerning state is D – uninterruptible sleep. Processes are in the uninterruptible sleep state if they actively wait for input/output operations to complete, so if there is a large number of such processes, it may mean that the input/output systems are overloaded.

Note that there are mysterious processes with command names in square brackets that you don't see in the output of `pstree`. Those are, in fact, kernel services that are made to look like processes for ease of monitoring, or for their own internal reasons.

If you want to also see what user owns each of those processes, you may want to add u to your command:

```
$ ps axu
USER          PID %CPU %MEM    VSZ    RSS TTY       STAT START    TIME
COMMAND
root          1   0.0  0.8 108084   8340 ?          Ss   Apr01    1:13 /usr/
lib/systemd/systemd --switched-root --system --deserialize 30
root          2   0.0  0.0      0      0 ?          S    Apr01    0:01
[kthreadd]
```

There are many more selection and formatting options, which you can find in the official documentation.

Process monitoring tools

The ps command gives you a static picture of processes, CPU and memory consumption. If you want to find out what causes CPU usage spikes, that's very inconvenient – you need to be lucky to run ps exactly when a spike occurs. That's why people wrote tools that monitor processes continuously and can display them sorted by resource consumption.

One of the oldest tools in this class is called top. It's widely available in Linux distribution repositories and may even be installed by default in your system. It displays an interactive process list, with processes that consume the largest amount of resources automatically floating to the top.

```
top - 16:24:25 up 150 days,  9:11,  1 user,  load average: 0.02, 0.01, 0.00
Tasks: 108 total,   2 running, 106 sleeping,   0 stopped,   0 zombie
%Cpu(s):  0.7 us,  0.7 sy,  0.0 ni, 98.3 id,  0.0 wa,  0.0 hi,  0.0 si,  0.3 st
MiB Mem :    959.1 total,    349.7 free,    200.9 used,    408.6 buff/cache
MiB Swap:    959.0 total,    837.2 free,    121.8 used.    621.9 avail Mem

    PID USER      PR  NI    VIRT    RES    SHR S  %CPU  %MEM     TIME+ COMMAND
    559 systemd+  20   0   17140    968    816 S   0.7   0.1 226:41.10 systemd-oomd
    799 mongod    20   0 1539624  20880  11644 S   0.7   2.1 934:31.55 mongod
 772712 dmbatur+  20   0   16300   5876   4392 S   0.3   0.6   0:00.24 sshd
      1 root      20   0  108084   8256   5252 S   0.0   0.8   1:14.04 systemd
      2 root      20   0       0      0      0 S   0.0   0.0   0:01.66 kthreadd
      3 root       0 -20       0      0      0 I   0.0   0.0   0:00.00 rcu_gp
      4 root       0 -20       0      0      0 I   0.0   0.0   0:00.00 rcu_par_gp
      6 root       0 -20       0      0      0 I   0.0   0.0   0:00.00 kworker/0:0H-events_highpri
      9 root       0 -20       0      0      0 I   0.0   0.0   0:00.00 mm_percpu_wq
     10 root      20   0       0      0      0 S   0.0   0.0   0:00.00 rcu_tasks_kthre
     11 root      20   0       0      0      0 S   0.0   0.0   0:00.00 rcu_tasks_rude_
     12 root      20   0       0      0      0 S   0.0   0.0   0:00.00 rcu_tasks_trace
     13 root      20   0       0      0      0 S   0.0   0.0   1:11.07 ksoftirqd/0
     14 root      20   0       0      0      0 R   0.0   0.0   8:48.55 rcu_preempt
     15 root      rt   0       0      0      0 S   0.0   0.0   0:35.26 migration/0
     16 root      20   0       0      0      0 S   0.0   0.0   0:00.00 cpuhp/0
     17 root      20   0       0      0      0 S   0.0   0.0   0:00.00 kdevtmpfs
     18 root       0 -20       0      0      0 I   0.0   0.0   0:00.00 netns
     19 root       0 -20       0      0      0 I   0.0   0.0   0:00.00 inet_frag_wq
     20 root      20   0       0      0      0 S   0.0   0.0   0:16.38 kauditd
     21 root      20   0       0      0      0 S   0.0   0.0   0:00.00 oom_reaper
     22 root       0 -20       0      0      0 I   0.0   0.0   0:00.00 writeback
     23 root      20   0       0      0      0 S   0.0   0.0   5:32.18 kcompactd0
     24 root      25   5       0      0      0 S   0.0   0.0   0:00.00 ksmd
     25 root      39  19       0      0      0 S   0.0   0.0   0:00.00 khugepaged
     26 root       0 -20       0      0      0 I   0.0   0.0   0:00.00 cryptd
     27 root       0 -20       0      0      0 I   0.0   0.0   0:00.00 kintegrityd
     28 root       0 -20       0      0      0 I   0.0   0.0   0:00.00 kblockd
```

Figure 4.1 – The top command output

There are other tools inspired by it, such as htop, which offers different user interfaces and additional functionality. There are also tools that monitor resource usage types that top or htop don't, such as iotop, a tool to monitor the input/output activity of processes.

The /proc filesystem

Finally, let's examine the lowest-level interface to process information – the /proc filesystem. It's an example of a *virtual filesystem* and a good illustration of the *everything is a file* principle widely used by Unix-like operating systems.

To users, /proc looks like a directory with files. Some of those files have nothing to do with processes and contain other system information instead – for example, /proc/version contains the running kernel version. Here's an example from a Fedora 35 system:

```
$ cat /proc/version
Linux version 5.16.18-200.fc35.x86_64 (mockbuild@bkernel01.iad2.
fedoraproject.org) (gcc (GCC) 11.2.1 20220127 (Red Hat 11.2.1-9), GNU
ld version 2.37-10.fc35) #1 SMP PREEMPT Mon Mar 28 14:10:07 UTC 2022
```

However, those files aren't files on disk, and not even the root user can write to them or delete them. The kernel simply makes them look like files so that users can use familiar tools to read them.

Whenever a process is launched, the kernel adds a subdirectory named /proc/<PID>/. For example, let's peek into the directory of the init process:

```
$ sudo ls /proc/1/
arch_status  auxv      cmdline      cpu_resctrl_groups  environ  fdinfo
latency   map_files  mountinfo    net      oom_adj      pagemap
projid_
map  schedstat  smaps      stat syscall      timers      wchan
attr         cgroup     comm        cpuset          exe      gid_map
limits    maps        mounts      ns        oom_score
patch_state    root      sessionid  smaps_rollup  statm    task
timerslack_ns
autogroup    clear_refs  coredump_filter  cwd            fd        io
loginuid  mem        mountstats  numa_maps  oom_score_
adj  personality
sched        setgroups  stack        status  timens_offsets  uid_map
```

That interface is too low-level for an end user most of the time. For example, environment variables passed to a process on launch can be found in a file named environ, but let's try to read it and see:

```
$ sudo cat /proc/1/environ
TERM=vt220BOOT_IMAGE=(hd0,gpt2)/vmlinuz-5.16.18-200.fc35.x86_64
```

That output appears to be garbled – variables have no separators. In reality, it's not a bug; that file simply contains the process environment data exactly as it appears in its memory, with variable pairs separated by non-printable null characters, as per the C programming language convention.

The same applies to /proc/1/cmdline – the file that contains a full command that was used to launch that process:

```
$ sudo cat /proc/1/cmdline
/usr/lib/systemd/systemd--switched-root--system--deserialize30
```

However, it's recognizable, and it's exactly where ps takes the process command; it just inserts spaces in place of null bytes to make it correct and readable:

```
$ ps ax | grep systemd
   1 ?      Ss      1:14 /usr/lib/systemd/systemd --switched-root
--system --deserialize 30
```

Thus, it's good to know about the raw /proc interface, but it's impractical to use it as a source of process information. Tools such as ps and pstree can present it in a much more readable form and allow you to filter it. However, it's also important to understand that those tools don't use any special kernel interfaces other than /proc, and in an emergency situation, you can always get the same information from there without any tools at all.

Summary

In this chapter, we learned that process startup is not a trivial operation, and even native code binaries are not simply loaded into memory byte by byte. We learned how to explore the process tree with the pstree command, how to force processes to terminate or reload with kill, and how to examine and interpret exit codes.

We also learned that the kernel communicates with running processes using POSIX signals, that different signals have different meanings, and that there are more signals than what the kill command allows the user to send. Apart from SIGTERM or SIGKILL, which are sent by users or userspace tools, there are many signals that the kernel uses to indicate programming errors and special conditions. Among them are SIGILL, for programs that attempt to execute illegal CPU instructions, and SIGPIPE, for cases when a connection is closed by the other side.

In the next chapter, we will learn how to discover and examine the hardware installed in a machine running Linux.

5

Hardware Discovery

Knowing how to find out what hardware an operating system runs on and what peripherals are attached is a necessary skill for all systems administrators – at the very least, every administrator needs to know how many CPU cores they have and how much memory there is to allocate resources to applications. In this chapter, we'll learn how to retrieve and interpret information about CPUs, USB peripherals, and storage devices, using both the raw kernel interfaces and utilities to work with them. We will also cover tools specific to x86 machines with a SMBIOS/DMI interface.

In this chapter, we will cover the following topics:

- How to discover the number of CPUs, their model names, and features
- How to discover PCI, USB, and SCSI peripherals
- How to use platform-specific tools to retrieve detailed hardware information from system firmware

Discovering CPU model and features

The central processor is certainly one of the most important hardware components, and there are many reasons to find out detailed information about it. The CPU model name or number and frequency are the first things you would look at to find out the age and overall performance of a machine. However, there are more details that are often useful in practice. For example, the number of CPU cores is important to know if you run applications that support multiple worker threads or processes (such as make -j2). Trying to run more processes than there are CPUs may slow the application down because some of those processes end up waiting for an available CPU, so you may want to run fewer worker processes to avoid overloading the machine.

It's also important to know whether your CPU supports specific acceleration technologies such as AES-NI or Intel QuickAssist. If they are available, some applications can perform much better if you enable support for those acceleration features.

Feature discovery on different platforms

One thing you should remember about CPU information discovery is that it's largely a feature of the CPU itself rather than a feature of Linux. If a CPU has no way to report certain information, it may be difficult or even impossible to figure it out from the software.

Many ARM and MIPS CPUs do not report their exact model names, so the only way to find them may be to just open the case and look at the chip.

For example, the popular Raspberry Pi 4 single-board computer uses an ARM Cortex A72 CPU, but it doesn't tell you that.

```
pi@raspberrypi4    :~ $ cat /proc/cpuinfo
processor          : 0
model name         : ARMv7 Processor rev 3 (v7l)
BogoMIPS           : 108.00
Features           : half thumb fastmult vfp edsp neon vfpv3 tls vfpv4
idiva idivt vfpd32 lpae evtstrm crc32
CPU implementer    : 0x41
CPU architecture   : 7
CPU variant        : 0x0
CPU part           : 0xd08
CPU revision       : 3
```

We will focus on x86 CPUs made by AMD and Intel because they are still the most common CPUs outside of embedded system markets, and they have the most extensive feature-reporting functionality.

The /proc/cpuinfo file

We discussed the /proc filesystem when we talked about process control in *Chapter 4, Processes and Process Control*. That filesystem is virtual – files under /proc aren't present on any disk, and the kernel simply presents that information as if it were stored in files, according to the *everything is a file* principle of the Unix philosophy.

Apart from information about running processes, the kernel also uses /proc for information about CPUs and memory. The file with CPU information is named /proc/cpuinfo, as we saw earlier.

Now, let's look at that file on an Intel machine:

```
$ cat /proc/cpuinfo
processor          : 0
vendor_id          : GenuineIntel
cpu family         : 6
model              : 44
model name         : Intel(R) Xeon(R) CPU           L5630  @ 2.13GHz
stepping           : 2
```

```
microcode          : 0x1a
cpu MHz            : 2128.000
cache size         : 12288 KB
physical id        : 0
siblings           : 1
core id            : 0
cpu cores          : 1
apicid             : 0
initial apicid     : 0
fpu                : yes
fpu_exception      : yes
cpuid level        : 11
wp                 : yes
flags              : fpu vme de pse tsc msr pae mce cx8 apic sep mtrr
pge mca cmov pat pse36 clflush dts mmx fxsr sse sse2 syscall nx rdtscp
lm constant_tsc arch_perfmon pebs bts nopl xtopology tsc_reliable
nonstop_tsc cpuid aperfmperf pni pclmulqdq ssse3 cx16 sse4_1 sse4_2
x2apic popcnt tsc_deadline_timer aes hypervisor lahf_lm pti tsc_adjust
dtherm ida arat
bugs               : cpu_meltdown spectre_v1 spectre_v2 spec_store_
bypass l1tf mds swapgs itlb_multihit
bogomips           : 4256.00
clflush size       : 64
cache_alignment    : 64
address sizes      : 42 bits physical, 48 bits virtual
power management   :

processor          : 1
vendor_id          : GenuineIntel
cpu family         : 6
model              : 44
model name         : Intel(R) Xeon(R) CPU          L5630  @ 2.13GHz
stepping           : 2
microcode          : 0x1a
cpu MHz            : 2128.000
cache size         : 12288 KB
physical id        : 2
siblings           : 1
core id            : 0
cpu cores          : 1
apicid             : 2
initial apicid     : 2
fpu                : yes
fpu_exception      : yes
cpuid level        : 11
```

```
wp                      : yes
flags                   : fpu vme de pse tsc msr pae mce cx8 apic sep mtrr
pge mca cmov pat pse36 clflush dts mmx fxsr sse sse2 syscall nx rdtscp
lm constant_tsc arch_perfmon pebs bts nopl xtopology tsc_reliable
nonstop_tsc cpuid aperfmperf pni pclmulqdq ssse3 cx16 sse4_1 sse4_2
x2apic popcnt tsc_deadline_timer aes hypervisor lahf_lm pti tsc_adjust
dtherm ida arat
bugs                    : cpu_meltdown spectre_v1 spectre_v2 spec_store_
bypass l1tf mds swapgs itlb_multihit
bogomips                : 4256.00
clflush size            : 64
cache_alignment         : 64
address sizes           : 42 bits physical, 48 bits virtual
power management        :
```

Some information is obvious and needs no interpretation. The model is Intel Xeon L5630, its frequency is 2.13 GHz, and it has 12 kilobytes of cache. The reason why CPU vendor names are written in seemingly strange ways such as GenuineIntel and AuthenticAMD is that, internally, the vendor string is stored as three 32-bit values, which gives 12 bytes. Note how both GenuineIntel and AuthenticAMD are 12 characters long – they were chosen to fill all 12 bytes with printable ASCII characters to avoid issues with different interpretations of null bytes on different systems.

One field of note is bogomips. Its name hints at the **Million Instructions per Second** (**MIPS**) performance metric, but that value isn't a useful indicator of overall performance. The Linux kernel uses it for internal calibration purposes, and you shouldn't use it to compare the performance of different CPUs.

The flags field has the highest information density and is the most difficult to interpret. A flag is simply a bit in the Flags register, and the meanings of those bits vary widely. Some of them aren't even set in the hardware CPU, such as the hypervisor flag, which indicates that the system runs on a virtual machine (but its absence doesn't mean anything, since not all hypervisors set it).

Many flags indicate whether a certain feature is present, although interpreting their nature and importance requires familiarity with the vendor's terminology. For example, the Neon feature from the Raspberry Pi output is a set of **Single Instruction, Multiple Data** (**SIMD**) instructions for ARM CPUs, comparable with Intel's SSE, which is often used in multimedia and scientific applications. When in doubt, consult the CPU vendor documentation.

Multi-processor systems

Most CPUs on the market today include multiple CPU cores, and server mainboards often have multiple sockets. The cpuinfo file includes all information necessary to figure out the CPU socket and core layout, but there are caveats.

The field you should look at to determine the number of CPU sockets is physical id. Note that physical id values aren't always consecutive, so you can't just look at the maximum value and should consider all the values present. You can find unique IDs by piping the cpuinfo file through the sort and uniq commands:

```
$ cat /proc/cpuinfo | grep "physical id" | sort | uniq
physical id    : 0
physical id    : 2
```

In virtual machines running on x86 hardware, all CPUs that the hypervisor presents to them will show up as if they were in different physical sockets. For example, if you have a hypervisor host with a single quad-core CPU and create a virtual machine with two virtual CPUs, it will look like two single-core CPUs in different sockets.

On many non-x86 architectures, such as ARM, all CPUs will look as if they are different physical CPUs, whether they are different chips or cores on the same chips.

On bare-metal x86 machines, you can find two cpuinfo entries with different physical IDs and examine them further to find out the number of cores in each socket.

Consider this cpuinfo entry from a Linux system running on a laptop:

```
processor          : 7
vendor_id          : GenuineIntel
cpu family         : 6
model              : 140
model name         : 11th Gen Intel(R) Core(TM) i5-1135G7 @ 2.40GHz
stepping           : 1
microcode          : 0xa4
cpu MHz            : 1029.707
cache size         : 8192 KB
physical id        : 0
siblings           : 8
core id            : 3
cpu cores          : 4
apicid             : 7
initial apicid     : 7
fpu                : yes
fpu_exception      : yes
cpuid level        : 27
wp                 : yes
flags              : fpu vme de pse tsc msr pae mce cx8 apic sep mtrr
pge mca cmov pat pse36 clflush dts acpi mmx fxsr sse sse2 ss ht tm pbe
syscall nx pdpe1gb rdtscp lm constant_tsc art arch_perfmon pebs bts
rep_good nopl xtopology nonstop_tsc cpuid aperfmperf tsc_known_freq
pni pclmulqdq dtes64 monitor ds_cpl vmx est tm2 ssse3 sdbg fma cx16
```

```
       xtpr pdcm pcid sse4_1 sse4_2 x2apic movbe popcnt tsc_deadline_timer
       aes xsave avx f16c rdrand lahf_lm abm 3dnowprefetch cpuid_fault epb
       cat_12 invpcid_single cdp_12 ssbd ibrs ibpb stibp ibrs_enhanced tpr_
       shadow vnmi flexpriority ept vpid ept_ad fsgsbase tsc_adjust bmi1
       avx2 smep bmi2 erms invpcid rdt_a avx512f avx512dq rdseed adx smap
       avx512ifma clflushopt clwb intel_pt avx512cd sha_ni avx512bw avx512vl
       xsaveopt xsavec xgetbv1 xsaves split_lock_detect dtherm ida arat pln
       pts hwp hwp_notify hwp_act_window hwp_epp hwp_pkg_req avx512vbmi umip
       pku ospke avx512_vbmi2 gfni vaes vpclmulqdq avx512_vnni avx512_bitalg
       avx512_vpopcntdq rdpid movdiri movdir64b fsrm avx512_vp2intersect md_
       clear ibt flush_l1d arch_capabilities
vmx flags     : vnmi preemption_timer posted_intr invvpid ept_x_only
       ept_ad ept_1gb flexpriority apicv tsc_offset vtpr mtf vapic ept vpid
       unrestricted_guest vapic_reg vid ple pml ept_mode_based_exec tsc_
       scaling
bugs          : spectre_v1 spectre_v2 spec_store_bypass swapgs
eibrs_pbrsb
bogomips      : 4838.40
clflush size  : 64
cache_alignment : 64
address sizes : 39 bits physical, 48 bits virtual
power management :
```

From the cpu cores field, you can see that it's a quad-core CPU.

Now, note that the entry has processor: 7 in it, and the siblings field is set to 8. It appears that the system has eight CPUs, even though it clearly has a single physical CPU chip (all entries have physical id: 0), and the reported number of cores in each entry is 4.

This is caused by *simultaneous multithreading* technologies such as AMD SMT and Intel Hyper-Threading. Those technologies allow a single CPU core to maintain the state of more than one execution thread, which can speed up certain applications. If every core supports two threads at once, to the operating system it looks as if the machine has twice as many CPU cores than it really does. For this reason, you can't determine the number of physical cores just by looking at the highest /proc/cpuinfo entry number, so you need to examine it more closely.

There is a utility named nproc in the GNU coreutils package whose purpose is to output the number of CPU cores in the system, and it can simplify your job if you need the number of CPUs – for example, to determine how many worker processes or threads of an application to spawn. However, it does not take simultaneous multithreading into account and prints the number of virtual cores if an SMT technology is enabled. If your application requires the number of physical cores, you should not rely on the output of nproc for that purpose. This is what the output of nproc would look like on that machine:

```
$ nproc
8
```

High-level CPU discovery utilities

As you can see, reading the /proc/cpuinfo file can be a tedious and tricky task. For this reason, people created utilities that aim to simplify it. The most popular tool in this category is lscpu from the util-linux package.

If offers an easier-to-read output:

```
$ lscpu
Architecture          : x86_64
CPU op-mode(s)        : 32-bit, 64-bit
Byte Order            : Little Endian
Address sizes         : 42 bits physical, 48 bits virtual
CPU(s)                : 2
On-line CPU(s) list   : 0,1
Thread(s) per core    : 1
Core(s) per socket    : 1
Socket(s)             : 2
NUMA node(s)          : 1
Vendor ID             : GenuineIntel
CPU family            : 6
Model                 : 44
Model name            : Intel(R) Xeon(R) CPU          L5630  @
2.13GHz
Stepping              : 2
CPU MHz               : 2128.000
BogoMIPS              : 4256.00
Hypervisor vendor     : VMware
Virtualization type   : full
L1d cache             : 32K
L1i cache             : 32K
L2 cache              : 256K
L3 cache              : 12288K
NUMA node0 CPU(s)     : 0,1
Flags                 : fpu vme de pse tsc msr pae mce cx8 apic sep
mtrr pge mca cmov pat pse36 clflush dts mmx fxsr sse sse2 syscall
nx rdtscp lm constant_tsc arch_perfmon pebs bts nopl xtopology tsc_
reliable nonstop_tsc cpuid aperfmperf pni pclmulqdq ssse3 cx16 sse4_1
sse4_2 x2apic popcnt tsc_deadline_timer aes hypervisor lahf_lm pti
tsc_adjust dtherm ida arat
```

Another advantage is that you can get a machine-readable JSON output by running lscpu --json and analyzing it with scripts.

Note, however, that lscpu doesn't take the simultaneous multithreading issue into account (at least as of version 2.38) and will report twice as many CPUs if AMD SMT or Intel Hyper-Threading is enabled.

As you can see, there is a lot of information about the CPUs that you can get either directly from the /proc filesystem or by using high-level utilities to simplify the process. However, you should always remember the nuances of interpretation, such as the issue of core numbers being inflated by simultaneous multithreading technologies.

Memory discovery

Discovering the amount of memory is often even more practically important than discovering CPU features. It is required to plan application deployment, choose the size of a swap partition, and estimate whether you need to install more memory already.

However, the kernel interfaces for memory discovery are not as rich as those for discovering CPU features. For example, it is impossible to find out how many memory slots a system has, how many of them are used, and what the sizes of memory sticks installed in those slots using the kernel interface are alone. At least on some architectures, it is possible to obtain that information, but from the firmware rather than from the kernel, as we will see later in the *dmidecode* section.

Moreover, information from the kernel can be misleading for beginners who are unfamiliar with Linux kernel conventions. First, let us look at that information and then discuss how to interpret it.

First, we will look at the output of the free utility that comes from the procps-ng package. That utility has the -m option to show memory amounts in megabytes rather than in kibibytes, which is much easier to read.

	total	used	free	shared	buff/cache	available
Mem:	15296	13849	596	47	850	1072
Swap:	38709	17865	20844			

Figure 5.1 – the free –m output

On the surface, its output seems self-descriptive. The total amount of memory here is about 16 gigabytes. Of that amount, 13849 megabytes are used. However, the question of how the kernel uses memory is not trivial. Not all memory that is declared as used is used by user applications – the kernel also uses memory to cache data from disk to speed up input-output operations. The more memory the system has that is not currently used by any processes, the more memory will be used by Linux for caching. Whenever there is not enough completely unused memory to allocate to a process that requests it, the kernel will evict some cached data to free up space. For this reason, the amount of memory in the used column can be very high, even on a system that runs very few applications, which may create an impression that the system is running out of memory when, in fact, there is lots of memory available to applications. The amount of memory used for disk caching is in the buff/cache column, and the amount of memory available to applications is in the available column.

Raw information about memory consumption is found in the /proc/meminfo file. The output of the free utility is a summary of information from that file.

Discovering PCI devices

Many peripherals are attached to a PCI bus. These days, this usually means **PCI Express** (**PCI-e**) rather than older PCI or PCI-x buses, but from a software point of view, they are all PCI devices, whichever variant of that bus they are attached to.

The kernel exposes information about them in the `/sys/class/pci_bus` hierarchy, but reading those files by hand would be a very time-consuming task, unlike in `/proc/cpuinfo`, so in practice, people always use utilities for it. The most popular one is `lspci` from the `pciutils` package.

Here is a sample output:

```
$ sudo lspci
00:00.0 Host bridge: Intel Corporation 11th Gen Core Processor Host
Bridge/DRAM Registers (rev 01)
00:02.0 VGA compatible controller: Intel Corporation TigerLake-LP GT2
[Iris Xe Graphics] (rev 01)
00:04.0 Signal processing controller: Intel Corporation TigerLake-LP
Dynamic Tuning Processor Participant (rev 01)
00:06.0 PCI bridge: Intel Corporation 11th Gen Core Processor PCIe
Controller (rev 01)
00:0a.0 Signal processing controller: Intel Corporation Tigerlake
Telemetry Aggregator Driver (rev 01)
00:14.0 USB controller: Intel Corporation Tiger Lake-LP USB 3.2 Gen
2x1 xHCI Host Controller (rev 20)
00:14.2 RAM memory: Intel Corporation Tiger Lake-LP Shared SRAM (rev
20)
00:14.3 Network controller: Intel Corporation Wi-Fi 6 AX201 (rev 20)
00:15.0 Serial bus controller: Intel Corporation Tiger Lake-LP Serial
IO I2C Controller #0 (rev 20)
00:15.1 Serial bus controller: Intel Corporation Tiger Lake-LP Serial
IO I2C Controller #1 (rev 20)
00:16.0 Communication controller: Intel Corporation Tiger Lake-LP
Management Engine Interface (rev 20)
00:17.0 SATA controller: Intel Corporation Tiger Lake-LP SATA
Controller (rev 20)
00:1c.0 PCI bridge: Intel Corporation Device a0bc (rev 20)
00:1d.0 PCI bridge: Intel Corporation Tiger Lake-LP PCI Express Root
Port #9 (rev 20)
00:1f.0 ISA bridge: Intel Corporation Tiger Lake-LP LPC Controller
(rev 20)
00:1f.4 SMBus: Intel Corporation Tiger Lake-LP SMBus Controller (rev
20)
00:1f.5 Serial bus controller: Intel Corporation Tiger Lake-LP SPI
Controller (rev 20)
```

```
01:00.0 3D controller: NVIDIA Corporation GP108M [GeForce MX330] (rev
a1)
02:00.0 Non-Volatile memory controller: Samsung Electronics Co Ltd
NVMe SSD Controller 980
03:00.0 Ethernet controller: Realtek Semiconductor Co., Ltd.
RTL8111/8168/8411 PCI Express Gigabit Ethernet Controller (rev 15)
```

Full access to the PCI bus information is only available to the `root` user, so you should run `lspci` with `sudo`. PCI device classes are standardized, so that utility will not only tell you about device vendors and models but also their functions, such as an Ethernet controller or a SATA controller.

Some of those names may be slightly misleading – for example, a VGA-compatible controller may not have an actual VGA port; these days, it's almost invariably DVI, HDMI, or Thunderbolt/DisplayPort.

Discovering USB devices

To discover USB devices, there's the `lsusb` utility from the `usbutils` package. That command does not require `root` privileges. Here's what its output may look like:

```
$ lsusb
Bus 002 Device 001: ID 1d6b:0003 Linux Foundation 3.0 root hub
Bus 001 Device 005: ID 1bcf:2b98 Sunplus Innovation Technology Inc.
Integrated_Webcam_HD
Bus 001 Device 036: ID 0b0e:0300 GN Netcom Jabra EVOLVE 20 MS
Bus 001 Device 035: ID 05ac:12a8 Apple, Inc. iPhone 5/5C/5S/6/SE
Bus 001 Device 006: ID 8087:0aaa Intel Corp. Bluetooth 9460/9560
Jefferson Peak (JfP)
Bus 001 Device 002: ID 047d:1020 Kensington Expert Mouse Trackball
Bus 001 Device 001: ID 1d6b:0002 Linux Foundation 2.0 root hub
```

Even though the USB bus specification also includes standardized device classes, `lsusb` does not show them by default. One reason for that behavior is that a single USB device may implement multiple functions. A smartphone, for example, can present itself as a mass storage device (similar to a USB stick) for generic file transfer and as a digital camera to retrieve pictures from it using **Picture Transfer Protocol (PTP)**.

You can get detailed information about a device by running `lsusb` with additional options. In the previous example, running `lsusb -s 2 -v` would show you information about device `002` – a Kensington trackball.

Note that if a device is shown in the output of `lsusb`, it doesn't always mean that it's attached to a USB port. Onboard devices may also be connected to the USB bus. In the previous example, the Bluetooth controller is an internal device in the laptop.

Discovering storage devices

Storage devices can be attached to different buses, so discovering them may be complicated. One useful command to discover everything that looks like a storage device is `lsblk` from the `util-linux` package (its name stands for list block devices):

```
$ lsblk
NAME     MAJ:MIN RM   SIZE RO TYPE MOUNTPOINTS
zram0    251:0     0  959M  0 disk [SWAP]
vda      252:0     0   25G  0 disk
├─vda1   252:1     0    1M  0 part
├─vda2   252:2     0  500M  0 part /boot
├─vda3   252:3     0  100M  0 part /boot/efi
├─vda4   252:4     0    4M  0 part
└─vda5   252:5     0 24.4G  0 part /home
                                   /
```

One caveat is that it will show virtual devices as well as physical ones. For example, if you mount an ISO image using a virtual loop device, it will show up as a storage device – because from the user's point of view, it is indeed a storage device:

```
$ sudo mount -t iso9660 -o ro,loop /tmp/some_image.iso /mnt/iso/
$ lsblk
NAME                                     MAJ:MIN RM   SIZE RO
TYPE   MOUNTPOINTS
loop0                                        7:0     0  1.7G  1
loop   /mnt/tmp
├─loop0p1                                  259:4     0   30K  1
part
└─loop0p2                                  259:5     0  4.3M  1
part
```

If you try to discover physical devices, you may want to try the `lsscsi` command from the package that is also usually called `lsscsi`:

```
$ lsscsi
[N:0:5:1]      disk      PM991a NVMe Samsung 512GB__1           /dev/
nvme0n1
```

A confusing part is that the *protocol* of the original parallel SCSI bus found many new applications and remains widely used, even though its original hardware implementation was replaced by newer buses. There are many devices that will show up in the output of that utility, including SATA, **Serial-Attached SCSI (SAS)**, and NVMe drives, as well as USB mass storage devices.

Conversely, paravirtual devices such as VirtIO drives in KVM and Xen virtual machines will not be included in the `lsscsi` output. All in all, you may need to rely on a combination of `lsusb`, `lsscsi`, and `lsblk` to get the full picture. A good thing is that none of those commands require `root` privileges. Another good thing about `lsblk` specifically is that you can run `lsblk --json` to get machine-readable output and load it in a script.

High-level discovery tools

As well as these tools, which are specific to a certain bus or device type, there are also tools that can help you discover all hardware present in a system.

dmidecode

On x86 systems, you can use a program named `dmidecode` to retrieve and view information from the firmware (BIOS/UEFI) via the Desktop Management Interface (hence its name). That interface is also known as SMBIOS. Since it's specific to the x86 PC firmware standards, it will not work on machines with other architectures such as ARM or MIPS, but on x86 laptops, workstations, and servers, it can help you discover information that cannot be obtained in any other way, such as the number of RAM slots.

One disadvantage is that you need `root` privileges to run it. Another thing to take note of is that it produces a lot of output. There is no way to reproduce a complete output in the book because it would take many pages, but for reference, here's what the beginning of its output may look like (in this case, in a VMware virtual machine):

```
$ sudo dmidecode
# dmidecode 3.2
Getting SMBIOS data from sysfs.
SMBIOS 2.4 present.
556 structures occupying 27938 bytes.
Table at 0x000E0010.

Handle 0x0000, DMI type 0, 24 bytes
BIOS Information
    Vendor: Phoenix Technologies LTD
    Version: 6.00
    Release Date: 09/21/2015
    Address: 0xE99E0
    Runtime Size: 91680 bytes
    ROM Size: 64 kB
    Characteristics:
        ISA is supported
        PCI is supported
```

```
            PC Card (PCMCIA) is supported
            PNP is supported
            APM is supported
            BIOS is upgradeable
            BIOS shadowing is allowed
            ESCD support is available
            Boot from CD is supported
            Selectable boot is supported
            EDD is supported
            Print screen service is supported (int 5h)
            8042 keyboard services are supported (int 9h)
            Serial services are supported (int 14h)
            Printer services are supported (int 17h)
            CGA/mono video services are supported (int 10h)
            ACPI is supported
            Smart battery is supported
            BIOS boot specification is supported
            Function key-initiated network boot is supported
            Targeted content distribution is supported
        BIOS Revision: 4.6
        Firmware Revision: 0.0

Handle 0x0001, DMI type 1, 27 bytes
System Information
        Manufacturer: VMware, Inc.
        Product Name: VMware Virtual Platform
        Version: None
        Serial Number: VMware-56 4d 47 fd 6f 6f 20 30-f8 85 d6 08 bf 09 fe
b5
        UUID: 564d47fd-6f6f-2030-f885-d608bf09feb5
        Wake-up Type: Power Switch
        SKU Number: Not Specified
        Family: Not Specified

Handle 0x0002, DMI type 2, 15 bytes
Base Board Information
        Manufacturer: Intel Corporation
        Product Name: 440BX Desktop Reference Platform
        Version: None
        Serial Number: None
        Asset Tag: Not Specified
        Features: None
        Location In Chassis: Not Specified
        Chassis Handle: 0x0000
```

```
    Type: Unknown
    Contained Object Handles: 0
```

As you can see in this output, there are exact model names for the chassis and the mainboard. Unlike, for example, a USB, where the protocol itself includes features that allow devices themselves to report their names and capabilities to the host, there is no cross-platform way to query the mainboard name, but at least on most x86 machines, that information is available through the DMI interface.

lshw

Another tool that provides comprehensive hardware reporting is `lshw` (LiSt HardWare). Like `dmidecode`, it can use the DMI interface as its information source, but it tries to support more hardware platforms and their platform-specific hardware discovery interfaces.

One disadvantage is that none of the popular distributions install it by default, so you'll always need to install it from the repositories by hand. It also requires `root` privileges to run, since it relies on the privileged access DMI.

Here is a sample output from a virtual machine running in KVM on the DigitalOcean cloud platform. Its output is also very large, so we only include its beginning:

```
$ sudo lshw
my-host
    description: Computer
    product: Droplet
    vendor: DigitalOcean
    version: 20171212
    serial: 293265963
    width: 64 bits
    capabilities: smbios-2.4 dmi-2.4 vsyscall32
    configuration: boot=normal family=DigitalOcean_Droplet
uuid=AE220510-4AD6-4B66-B49F-9D103C60BA5A
  *-core
      description: Motherboard
      physical id: 0
    *-firmware
        description: BIOS
        vendor: DigitalOcean
        physical id: 0
        version: 20171212
        date: 12/12/2017
        size: 96KiB
    *-cpu
        description: CPU
```

```
        product: DO-Regular
        vendor: Intel Corp.
        physical id: 401
        bus info: cpu@0
        slot: CPU 1
        size: 2GHz
        capacity: 2GHz
        width: 64 bits
        capabilities: fpu fpu_exception wp vme de pse tsc msr pae
mce cx8 apic sep mtrr pge mca cmov pat pse36 clflush mmx fxsr sse
sse2 ss syscall nx rdtscp x86-64 constant_tsc arch_perfmon rep_good
nopl cpuid tsc_known_freq pni pclmulqdq vmx ssse3 fma cx16 pcid sse4_1
sse4_2 x2apic movbe popcnt tsc_deadline_timer aes xsave avx f16c
rdrand hypervisor lahf_lm abm cpuid_fault invpcid_single pti ssbd ibrs
ibpb tpr_shadow vnmi flexpriority ept vpid ept_ad fsgsbase tsc_adjust
bmi1 avx2 smep bmi2 erms invpcid xsaveopt md_clear
   *-memory
        description: System Memory
        physical id: 1000
        size: 1GiB
        capacity: 1GiB
        capabilities: ecc
        configuration: errordetection=multi-bit-ecc
      *-bank
           description: DIMM RAM
           physical id: 0
           slot: DIMM 0
           size: 1GiB
           width: 64 bits
    *-pci
        description: Host bridge
        product: 440FX - 82441FX PMC [Natoma]
        vendor: Intel Corporation
        physical id: 100
        bus info: pci@0000:00:00.0
        version: 02
        width: 32 bits
        clock: 33MHz
      *-isa
           description: ISA bridge
           product: 82371SB PIIX3 ISA [Natoma/Triton II]
           vendor: Intel Corporation
           physical id: 1
           bus info: pci@0000:00:01.0
           version: 00
```

```
             width: 32 bits
             clock: 33MHz
             capabilities: isa
             configuration: latency=0
       *-ide
             description: IDE interface
             product: 82371SB PIIX3 IDE [Natoma/Triton II]
             vendor: Intel Corporation
             physical id: 1.1
             bus info: pci@0000:00:01.1
             version: 00
             width: 32 bits
             clock: 33MHz
             capabilities: ide isa_compat_mode bus_master
             configuration: driver=ata_piix latency=0
             resources: irq:0 ioport:1f0(size=8) ioport:3f6
   ioport:170(size=8) ioport:376 ioport:c160(size=16)
        *-usb
             description: USB controller
             product: 82371SB PIIX3 USB [Natoma/Triton II]
             vendor: Intel Corporation
             physical id: 1.2
             bus info: pci@0000:00:01.2
             version: 01
             width: 32 bits
             clock: 33MHz
             capabilities: uhci bus_master
             configuration: driver=uhci_hcd latency=0
             resources: irq:11 ioport:c0c0(size=32)
        *-usbhost
                 product: UHCI Host Controller
                 vendor: Linux 5.16.18-200.fc35.x86_64 uhci_hcd
                 physical id: 1
                 bus info: usb@1
                 logical name: usb1
                 version: 5.16
                 capabilities: usb-1.10
                 configuration: driver=hub slots=2 speed=12Mbit/s
```

As you can see, the output includes information that can only be retrieved from DMI (such as the mainboard model), together with information that we have seen in the outputs of lsusb and other utilities. In this sense, lshw can replace them if you want a detailed and complete overview of all installed hardware.

Summary

In this chapter, we learned that the Linux kernel can gather a lot of information about the system hardware and provide it to the user. In an emergency situation, it's possible to retrieve all that information directly from the kernel, using the /proc and /sys filesystems and reading files such as /proc/cpuinfo.

However, high-level utilities such as lscpu, lsscsi, and lsusb can make it much easier to retrieve information and analyze it.

There are also platform-specific utilities, such as dmidecode for x86 PCs, that can help you retrieve even more detailed information that cannot be retrieved otherwise, such as the number of memory slots.

In the next chapter, we will learn about configuring basic system settings.

Part 2:
Configuring and Modifying Linux Systems

The second part of the book is dedicated to managing individual Linux systems. Once you are comfortable interacting with the system, the next step is to learn how to configure it. In this part, you will learn how to configure basic settings such as the system hostname, how to create and manage users and groups, how to install additional software from package files or remote repositories, and how to set up and debug network connections and storage devices.

This part has the following chapters:

- *Chapter 6, Basic System Settings*
- *Chapter 7, User and Group Management*
- *Chapter 8, Software Installation and Package Repositories*
- *Chapter 9, Network Configuration and Troubleshooting*
- *Chapter 10, Storage Management*

6
Basic System Settings

Linux is a highly customizable operating system, and it provides a vast array of configuration options that allow users to tailor their systems to their specific needs. In this chapter, we will explore some of the basic system configuration settings in Linux and how they can be customized to improve system performance, security, and usability.

Before you make any changes to a configuration file, you should always make a backup. When making a copy, append a `.bak` extension to it so that you know it's a copy meant for safekeeping.

Eventually, you are almost guaranteed to make a blunder when modifying these files. It is essential to have a backup of any configuration files before making any changes.

Due to the impossibility of covering every configuration file in Linux, we will focus on the most common configurations instead:

- The `hosts` configuration file
- The `resolv` configuration file
- The `network-scripts` configuration file
- The `dhclient` configuration file
- The `sysctl` configuration file

Overview of basic settings

Linux has various basic settings that you can configure to customize the behavior of your system. These settings are typically found in configuration files, and they can affect various aspects of the operating system:

- **System time configuration**: The system time in Linux is critical for a variety of tasks, including scheduling tasks, logging, and time-sensitive applications. The system time can be configured using the `timedatectl` command in most modern Linux distributions. This command allows users to set the system's time zone, as well as the date and time.

- **Hostname configuration:** The hostname is the name given to a computer or device on a network. In Linux, the hostname can be configured using the `hostnamectl` command. This command allows users to set the hostname, as well as the static IP address and domain name.

- **User and group configuration:** In Linux, users and groups are used to control access to the system and its resources. The `useradd` and `groupadd` commands are used to create new users and groups, respectively. The `usermod` and `groupmod` commands are used to modify existing users and groups.

- **Network configuration:** Networking is an essential component of modern computing, and Linux provides several tools to configure network settings. The `ifconfig` command can be used to configure network interfaces, while the `ip` command can be used to manage IP addresses, routes, and network devices. The **NetworkManager** service is a popular tool for managing network connections and settings in Linux.

- **System security configuration:** Linux is known for its robust security features, and many system configuration settings focus on enhancing system security. Some of the key security configuration settings include configuring the firewall, managing user permissions, configuring **Security-Enhanced Linux (SELinux)**, and setting up system auditing and monitoring.

- **System performance configuration:** Linux is a highly efficient operating system, but there are still several configuration settings that can be used to optimize system performance. These settings include configuring the kernel, tuning system parameters such as the I/O scheduler and memory allocation, and managing system resources such as CPU and memory usage:

  - **Kernel parameters:** The Linux kernel has various tunable parameters that can be adjusted to optimize performance for specific workloads. These parameters can be set during boot or runtime using the `sysctl` command or by modifying the `/etc/sysctl.conf` file.

    For example, to increase the maximum number of open files allowed by the system, you can set the `fs.file-max` parameter:

    ```
    # Increase the maximum number of open files
    sysctl -w fs.file-max=100000
    ```

  - **CPU scaling:** Linux provides CPU scaling mechanisms that control the CPU's frequency and power-saving features. Adjusting CPU scaling can help strike a balance between power efficiency and performance.

    For example, to set the CPU governor to performance mode (maximum frequency all the time), you can use the `cpufreq-set` command (may require installation of `cpufrequtils`):

    ```
    cpufreq-set -r -g performance
    ```

  - **I/O scheduler:** The Linux kernel offers multiple I/O schedulers, each designed for specific storage devices and access patterns. Choosing the right scheduler for your storage can improve I/O performance.

As an example, to set the I/O scheduler for a specific block device, such as an SSD, use the following command:

```
echo "deadline" > /sys/block/sdX/queue/scheduler
```

- **Swap configuration**: The swap space is used when physical RAM is full. However, excessive swapping can significantly impact performance. Adjusting swapiness can control the tendency to swap out memory.

  For example, to reduce swappiness (less aggressive swapping), set a lower value (for example, 10) in /etc/sysctl.conf:

  ```
  vm.swappiness=10
  ```

- **Filesystem mount options**: Mount options for filesystems can impact performance based on the use case. Some options can optimize read/write operations or enhance data safety.

  As an example, for an SSD-mounted filesystem, you can set the noatime option to avoid updating access timestamps for improved read performance:

  ```
  UUID=YOUR_UUID /mnt/ssd ext4 defaults,noatime 0 2
  ```

- **Network settings**: Fine-tuning network parameters can enhance networking performance and reduce latency.

  For example, to increase the TCP buffer size, set the following in /etc/sysctl.conf:

  ```
  net.core.rmem_max = 16777216
  net.core.wmem_max = 16777216
  ```

System performance configuration is an iterative and adaptive process that requires a deep understanding of the system's components and workloads. By making informed and measured adjustments, system administrators can create efficient, stable, and responsive systems that meet the needs of their users and applications.

> **Note**
>
> Remember to back up configuration files before making changes and thoroughly test the system after modifications to ensure the desired performance improvements. The optimal settings may vary depending on the specific hardware, workload, and user requirements. Regular monitoring and profiling can help you identify performance bottlenecks and further refine the configuration.

The hosts configuration file

The Linux hosts file is a simple text file that is used to map hostnames to IP addresses. It is located in the /etc directory. The file contains a list of IP addresses and their corresponding hostnames. When a user tries to access a hostname, the system checks the hosts file to determine the IP address associated with that hostname.

The hosts file is used first in the **Domain Name System** (**DNS**) server's resolution process. If a hostname is not found in the DNS server, the system checks the hosts file for a mapping. This can be useful in situations where a specific hostname needs to be redirected to a different IP address or when you want to test a website before it is made live.

The hosts file consists of several columns of data, separated by whitespace. The first column contains the IP address, while the second column contains the hostname. Additional columns can be used to specify aliases for the hostname. For example, the following line in the hosts file maps the 192.168.1.80 IP address to the centos hostname and sets centos8 as an alias:

```
[~ $more /etc/hosts
127.0.0.1          localhost
127.0.1.1          ubuntu22
192.168.1.80     centos   centos8
```

Figure 6.1 – The content of the hosts file

Now, for example, if you want to SSH into that server with CentOS (192.168.1.80), you don't have to remember the IP, you just have to know what name you assigned to that server.

This is valid for ping too. You just have to remember the name when you use ping rather than the IP. Here's an example:

```
ping centos8
```

You can edit the hosts file using a text editor, such as nano or Vim. It's important to note that the hosts file requires root privileges to edit. Therefore, you must use the sudo command or switch to the root user before editing the file.

In addition to mapping hostnames to IP addresses, the hosts file can be used to block access to specific websites. By mapping a hostname to the loopback address (127.0.0.1), you can effectively block access to that website. For example, the following line in the hosts file blocks access to the example.com website:

```
127.0.0.1 example.com
```

In conclusion, the Linux hosts file is a simple yet powerful tool for mapping hostnames to IP addresses and overriding the DNS resolution process. It can be used to redirect traffic to a different IP address, test websites before they go live, and block access to specific websites. Understanding how to use the hosts file can help you troubleshoot networking issues and improve the security of your system.

The resolv configuration file

The `resolv.conf` file is an essential configuration file in Linux systems that is used to configure DNS resolution settings. DNS is responsible for translating human-readable domain names (such as `www.example.com`) into IP addresses that computers can understand. This translation is crucial for accessing websites, services, and other network resources on the internet. The file is located in the `/etc` directory and contains information about the DNS servers that the system should use to resolve domain names.

The `resolv.conf` file is used by various networking programs, including the system resolver library, web browsers, and email clients. When a user tries to access a website or connect to a remote server using its hostname, the system uses the `resolv.conf` file to find the corresponding IP address.

The `resolv.conf` file consists of several directives that define the DNS servers to use and the search domains to append to hostnames. The following are the most commonly used directives in the `resolv.conf` file:

- `nameserver`: This directive specifies the IP address of the DNS server to use for name resolution. You can specify multiple nameservers in the file, and the system will use them in the order they are listed. For example, the following line in the `resolv.conf` file specifies the IP address of the DNS server at `192.168.1.1`:

```
~ $more /etc/resolv.conf
# Generated by NetworkManager
nameserver 192.168.1.1
~ $
```

Figure 6.2 – The content of the resolv.conf file

- `search`: This directive specifies the search domain to append to hostnames that are not fully qualified. For example, if the search domain is set to `example.com`, and a user tries to access the www hostname, the system will try to resolve `www.example.com`. You can specify multiple search domains in the file, and the system will use them in the order they are listed. For example, the following line in the `resolv.conf` file specifies the search domain as `example.com`:

  search example.com

- `options`: This directive specifies additional options to use for name resolution, such as the timeout and retry values. For example, the following line in the `resolv.conf` file specifies a timeout of 5 seconds and a retry value of 2:

  options timeout:5 retries:2

You can edit the `resolv.conf` file using a text editor, such as nano or Vim. We need root privilege to edit. It's important to note that the `resolv.conf` file can be automatically generated by various tools, such as the **Dynamic Host Configuration Protocol** (**DHCP**) client or the NetworkManager service. Therefore, any changes you make to the file may be overwritten by these tools.

In conclusion, the Linux `resolv.conf` file is a crucial configuration file that defines the DNS servers to use and the search domains to append to hostnames. Understanding how to configure the `resolv.conf` file can help you troubleshoot networking issues and improve the performance and security of your system.

The network-scripts configuration file

In Linux, `network-scripts` is used to configure network interfaces. These scripts are located in the `/etc/sysconfig/network-scripts` directory and define the network settings for each interface, such as the IP address, netmask, gateway, and DNS servers (network file configuration path is specific to CentOS).

The `network-scripts` is written in Bash and consists of several configuration files, each corresponding to a specific network interface. The most commonly used configuration files are `ifcfg-ethX` for Ethernet interfaces and `ifcfg-wlanX` for wireless interfaces, where X is the interface number.

The `ifcfg-ethX` configuration file contains the following parameters:

- `DEVICE`: This parameter specifies the name of the network interface – for example, `DEVICE=eth0`.
- `BOOTPROTO`: This parameter specifies whether the interface should use DHCP or a static IP address. If DHCP is used, the parameter's value is set to `dhcp`. If a static IP address is used, the parameter's value is set to `static`.
- `IPADDR`: This parameter specifies the IP address of the interface.
- `NETMASK`: This parameter specifies the network mask of the interface.
- `GATEWAY`: This parameter specifies the default gateway for the interface.
- `DNS1`, `DNS2`, and `DNS3`: These parameters specify the IP addresses of the DNS servers to use for name resolution.

The `ifcfg-wlanX` configuration file contains similar parameters but also includes additional parameters for wireless settings, such as the ESSID and encryption method.

You can edit the `network-scripts` configuration files using a text editor, such as nano or Vim. It's important to note that changes made to the configuration files will not take effect until the network service is restarted. You can restart the network service by running the `service network restart` or `systemctl restart network` command, depending on your Linux distribution.

In addition to the configuration files, the `network-scripts` directory also contains scripts that are executed when the network service is started or stopped. These scripts can be used to perform additional network configuration tasks, such as setting up virtual interfaces or configuring network bridges.

In conclusion, the Linux `network-scripts` configuration files are used to configure network interfaces and define network settings such as IP addresses, netmasks, gateways, and DNS servers. Understanding how to configure these files can help you troubleshoot networking issues and improve the performance and security of your system.

An example of the `network-scripts` config file from CentOS that's been generated automatically is as follows:

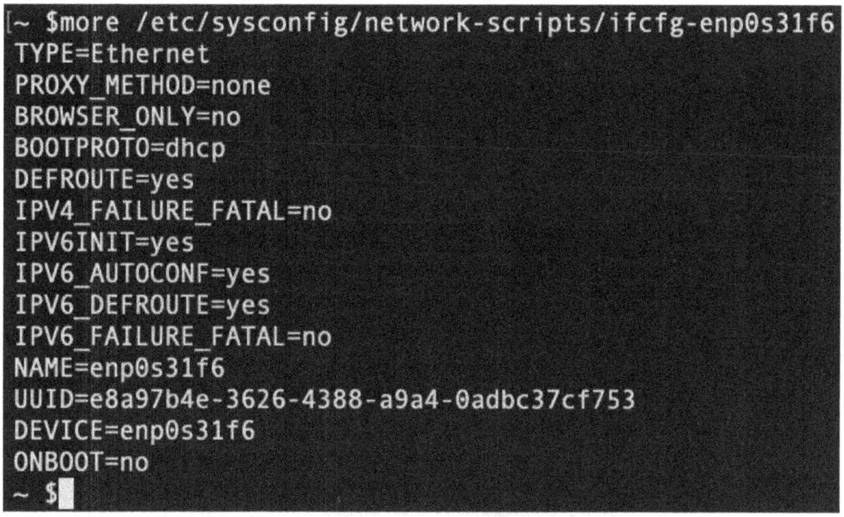

```
[~ $more /etc/sysconfig/network-scripts/ifcfg-enp0s31f6
TYPE=Ethernet
PROXY_METHOD=none
BROWSER_ONLY=no
BOOTPROTO=dhcp
DEFROUTE=yes
IPV4_FAILURE_FATAL=no
IPV6INIT=yes
IPV6_AUTOCONF=yes
IPV6_DEFROUTE=yes
IPV6_FAILURE_FATAL=no
NAME=enp0s31f6
UUID=e8a97b4e-3626-4388-a9a4-0adbc37cf753
DEVICE=enp0s31f6
ONBOOT=no
~ $
```

Figure 6.3 – Network configuration file

As you can see, DHCP is enabled and there is no static reservation, and the name of the network card is `enp0s31f6`.

If we would like to make a static IP reservation, we should use something like this:

```
HWADDR=$SOMETHING
TYPE=Ethernet
BOOTPROTO=none // turns off DHCP
IPADDR=192.168.1.121 // set your IP
PREFIX=24 // subnet mask
GATEWAY=192.168.1.1
DNS1=1.1.1.2 // set your own DNS
DNS2=8.8.8.8
DNS3=8.8.4.4
DEFROUTE=yes
IPV4_FAILURE_FATAL=no
NAME=enp0s31f6
DEVICE=enp0s31f6
ONBOOT=yes // starts on boot
```

With Ubuntu 17, networking configuration is done through Netplan, a YAML-based framework. You can do all of your DNS, gateway, netmask, and IP configuration here.

/etc/netplan is where the network configuration files are located.

A sample configuration file for a network interface using Netplan looks like this:

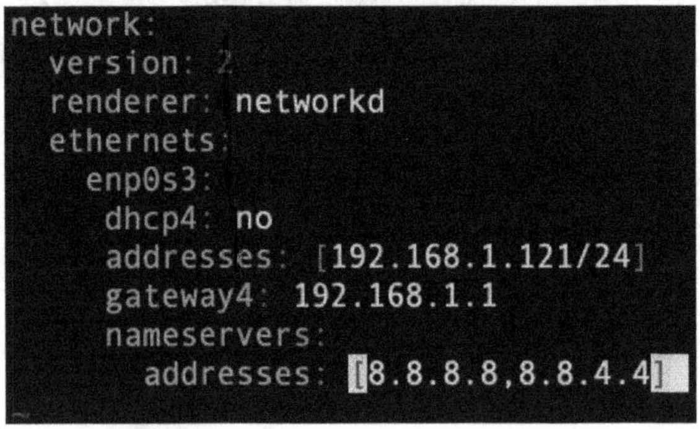

```
network:
  version: 2
  renderer: networkd
  ethernets:
    enp0s3:
    dhcp4: no
    addresses: [192.168.1.121/24]
    gateway4: 192.168.1.1
    nameservers:
      addresses: [8.8.8.8,8.8.4.4]
```

Figure 6.4 – Ubuntu Netplan configuration file

To have your changes take effect, restart networking with the netplan apply command.

The dhclient configuration file

The dhclient configuration file in Linux is located at /etc/dhcp/dhclient.conf. This file is used by the dhclient program to configure the DHCP client settings on a Linux system.

The dhclient.conf file contains various configuration options that control how the DHCP client interacts with the DHCP server. Some of the common configuration options that can be set in the dhclient.conf file are as follows:

- **Timeout values**: The dhclient.conf file allows you to set the timeout values for various DHCP requests. You can set the timeout values for DHCPDISCOVER, DHCPREQUEST, and DHCPACK messages.

- **Lease time**: You can set the length of time that the DHCP client can use an assigned IP address.

- **DNS servers**: The dhclient.conf file allows you to specify the DNS servers that the DHCP client should use.

- **Hostname**: You can set the hostname that the DHCP client should use when requesting an IP address.

- **Client identifier**: You can specify the client identifier that the DHCP client should use when communicating with the DHCP server.

- **Vendor-specific options**: The dhclient.conf file also allows you to set vendor-specific options that can be used to configure various network settings.

It's important to note that the syntax for the dhclient.conf file can vary, depending on the version of dhclient that is being used. It's always a good idea to consult the documentation for your specific version of dhclient to ensure that you are using the correct syntax.

Here is what a default `dhclient.conf` file will look like in Ubuntu:

```
~ $more /etc/dhcp/dhclient.conf
# Configuration file for /sbin/dhclient.
#
# This is a sample configuration file for dhclient. See dhclient.conf's
#       man page for more information about the syntax of this file
#       and a more comprehensive list of the parameters understood by
#       dhclient.
#
# Normally, if the DHCP server provides reasonable information and does
#       not leave anything out (like the domain name, for example), then
#       few changes must be made to this file, if any.
#

option rfc3442-classless-static-routes code 121 = array of unsigned integer 8;

send host-name = gethostname();
request subnet-mask, broadcast-address, time-offset, routers,
        domain-name, domain-name-servers, domain-search, host-name,
        dhcp6.name-servers, dhcp6.domain-search, dhcp6.fqdn, dhcp6.sntp-servers,
        netbios-name-servers, netbios-scope, interface-mtu,
        rfc3442-classless-static-routes, ntp-servers;

#send dhcp-client-identifier 1:0:a0:24:ab:fb:9c;
#send dhcp-lease-time 3600;
#supersede domain-name "fugue.com home.vix.com";
#prepend domain-name-servers 127.0.0.1;
#require subnet-mask, domain-name-servers;
timeout 300;
#retry 60;
#reboot 10;
#select-timeout 5;
#initial-interval 2;
#script "/sbin/dhclient-script";
#media "-link0 -link1 -link2", "link0 link1";
#reject 192.33.137.209;

#alias {
#   interface "eth0";
#   fixed-address 192.5.5.213;
#   option subnet-mask 255.255.255.255;
#}

#lease {
#   interface "eth0";
#   fixed-address 192.33.137.200;
#   medium "link0 link1";
#   option host-name "andare.swiftmedia.com";
#   option subnet-mask 255.255.255.0;
#   option broadcast-address 192.33.137.255;
#   option routers 192.33.137.250;
#   option domain-name-servers 127.0.0.1;
#   renew 2 2000/1/12 00:00:01;
#   rebind 2 2000/1/12 00:00:01;
#   expire 2 2000/1/12 00:00:01;
#}
~ $
```

Figure 6.5 – DHCP configuration file

In conclusion, the dhclient.conf configuration file plays a crucial role in managing the DHCP client behavior on a system. By customizing the dhclient.conf file, administrators can fine-tune various settings and options to ensure optimal network connectivity and address assignment.

Properly configuring dhclient.conf can greatly enhance network stability, security, and performance. For example, administrators can prioritize certain DNS servers for faster resolution, enforce specific lease durations to manage IP address allocation efficiently, and set up fallback mechanisms in case the primary DHCP server becomes unavailable.

The sysctl configuration file

The Linux sysctl configuration file is located at /etc/sysctl.conf. This file is used to configure kernel parameters at runtime. The sysctl.conf file contains a set of key-value pairs that represent various kernel parameters.

The sysctl.conf file is divided into sections, where each section contains a set of key-value pairs that correspond to a specific group of kernel parameters. Each key-value pair consists of the name of the kernel parameter, followed by its value. This value can be either a numeric value or a string.

Here are some examples of kernel parameters that can be configured using the sysctl.conf file:

- net.ipv4.ip_forward: This parameter enables or disables IP forwarding. A value of 1 enables IP forwarding, while a value of 0 disables it.
- net.ipv4.tcp_syncookies: This parameter enables or disables TCP SYN cookies. A value of 1 enables TCP SYN cookies, while a value of 0 disables them.
- kernel.core_pattern: This parameter specifies the pattern used to name core dump files. The default pattern is core, but you can specify a different pattern if you wish.
- kernel.shmmax: This parameter specifies the maximum size of a shared memory segment in bytes.
- vm.swappiness: This parameter controls the degree to which the kernel swaps out unused memory pages to disk. A higher value means that the kernel will be more aggressive in swapping out memory pages, while a lower value means that the kernel will be less aggressive.

To apply changes made to the sysctl.conf file, you can use the sysctl command with the -p option, which will load the settings from the file into the kernel. It's important to note that some kernel parameters can only be set at boot time and cannot be changed at runtime using the sysctl.conf file.

Depending on your system and kernel, the sysctl.conf file might be called or located in the following locations:

- /etc/sysctl.d/*.conf
- /run/sysctl.d/*.conf

- `/usr/local/lib/sysctl.d/*.conf`
- `/usr/lib/sysctl.d/*.conf`
- `/lib/sysctl.d/*.conf`
- `/etc/sysctl.conf`

Linux will often try to read those files in that order. The remaining files with the same name in other folders are ignored the first time it discovers a genuine file with valid entries.

If you don't know what you're doing, then playing around with the `etc/sysctl.conf` file could have serious consequences. You can observe which commands and files the OS tries to load and in what order by running the `sysctl --system` command. However, use this command with caution. This command will actively load and conduct operations on your OS because it is not a dry-run command, and you run the risk of misconfiguring your settings if you're unsure if it should do that.

Summary

In conclusion, Linux provides a vast array of configuration settings that can be used to tailor the system to specific needs. Understanding how to configure basic system settings such as system time, hostname, user and group settings, network settings, security settings, and performance settings can help improve system performance, security, and usability. By customizing these settings, users can create a Linux system that meets their specific requirements and enhances their productivity.

In the next chapter, we will talk about user and group management.

7

User and Group Management

If you administer Linux servers, the users of such servers can be both your greatest asset and your greatest headache. Throughout the course of your career, you will be responsible for the addition of a large number of new users, the management of their passwords, the deletion of their accounts when they leave the organization, and the provision and revocation of access to resources located across the network. Even on servers where you are the sole user, you will still be responsible for managing user accounts. This is because even system processes run under the guise of a user. If you want to be effective at managing Linux servers, you will also need to know how to manage rights, implement password policies, and limit the number of people who can execute administrative commands on the computer. In this chapter, we'll go over these topics in detail so that you have a solid understanding of how to manage users and the resources they consume.

In this chapter, we're going to learn how to do the following:

- Create users and groups
- Modify users and groups
- Delete users and groups
- List all users
- Add a user to a group
- Remove a user from a group

Overview of managing accounts/groups

As a multi-user operating system, Linux allows for multiple users to be logged in and work simultaneously on a single machine. Be aware that it is never a good idea to let users share login information for the same account. It is preferable to have as many accounts as users who require machine access.

Access to specific system resources, such as directories and files, may need to be shared by two or more users. We can achieve both goals using Linux's user and group administration features.

General/normal users and root/superusers are the two categories of users in Linux systems.

One of the fundamental components of the Linux operating system is the management of user and group accounts. The custom rights of user and group accounts are maintained by each user logging in to the operating system using a different set of credentials. Adding new users requires specific permissions (superuser); the same holds true for other user or group administration operations, including account deletion, account update, and group addition and deletion.

These operations are performed using the following commands:

- `adduser`: Add a user to the system
- `userdel`: Delete a user account and related files
- `addgroup`: Add a group to the system
- `delgroup`: Remove a group from the system
- `usermod`: Modify a user account
- `chage`: This command is used to change the password expiration time and see user password expiry information
- `passwd`: This command is used to create or change a user account's password
- `sudo`: Run one or more commands as another user (typically with superuser permissions by running the `sudo su <username>` command)

Files relevant to these operations include `/etc/passwd` (user information), `/etc/shadow` (encrypted passwords), `/etc/group` (group information), and `/etc/sudoers` (configuration for `sudo`).

Superuser access is granted by using either the `su` command to become the root user or the `sudo su` command to get root privileges.

These are the default locations for user account information:

- User account properties: `/etc/passwd`
- User password properties: `/etc/shadow`

A group with the same username is also created when a user is created. Every user has a home directory; for the root user, it is placed in `/root`; for all other users, it is in `/home/`.

The `/etc/passwd` file contains all of the account details. This file has the following structure and includes a record for each system user account (fields are delimited by colons):

```
<username>:<x>:<UID>:<GID>:<Comment>:<Home directory>:<Default shell>
```

Let's carefully examine the preceding code:

- The <username> and <Comment> fields are self-explanatory

- The shadowed password (in /etc/shadow), which is required to log on as <username>, is indicated by the x in the second field

- The <UID> and <GID> entries include integers that, respectively, reflect the <usernameprimary> Group ID and User ID

- <Home directory> displays the full path to the home directory of the current user

- When a user logs in to the system, <Default shell> is the shell that is made available to them

Now that we have an overview of the commands used to manage accounts/groups, let's start to play with them.

How to add a new account

There are two commands, adduser and useradd, that can be utilized in Linux for the purpose of generating new users. The fact that these two commands achieve the same thing (albeit in different ways) and have names that are extremely similar to one another can make this a bit difficult to understand at first. I'll begin by walking you through the useradd command, and then I'll describe how adduser works differently. It's possible that you will prefer the latter option, but we'll discuss that in a moment.

Using useradd

You need sudo capabilities in order to add an account if you don't have root access. This must be defined in /etc/sudoers.

To begin, here is a working example of the useradd command that you can put to use:

```
sudo  useradd -d /home/packt -m packt
```

I set up a new user with the name packt by using this command. I am confirming that I want a home directory to be established for this user by using the -d option, and after that, I specified /home/ packt as the user's home directory in the following command. If I hadn't used the -m parameter, the system would not have known that I wanted my home directory to be created while the process was running; in that case, I would have had to manually create the directory. At the end, I announced the username that will be used for my new user (in this case, packt). As we progress through this book, we will come across commands that, in order to be executed, need root privileges. This was demonstrated in the preceding command. I'll prefix commands that need certain permissions with sudo, as that is the standard way to accomplish it. When you see this, it simply indicates that the command can only be executed with the root user's rights. You can also execute these commands by logging in as root (if root access is enabled) or switching to root mode. Both of these options are available to you. However, rather than logging in with the root account, it is strongly recommended that you use sudo instead. I explained this earlier.

The following command can be used to set a password for the newly created packt account:

```
~ $sudo passwd packt
Changing password for user packt.
New password:
Retype new password:
passwd: all authentication tokens updated successfully.
```

Another command is adduser, which will create everything automatically for you, including the home directory and group, and will ask you to set the password.

Using adduser

The adduser command is yet another option for establishing a user account, as mentioned earlier. When you first use this command, the difference it makes (as well as the convenience it provides) should become instantly evident. Execute the adduser command while providing the username of the new user you want to create. Go ahead and give it a shot:

```
~ $sudo adduser packt2
Adding user `packt2' ...
Adding new group `packt2' (1004) ...
Adding new user `packt2' (1003) with group `packt2' ...
The home directory `/home/packt2' already exists.  Not copying from `/
etc/skel'.
New password:
Retype new password:
passwd: password updated successfully
Changing the user information for packt2
Enter the new value, or press ENTER for the default
        Full Name []:
        Room Number []:
        Work Phone []:
        Home Phone []:
        Other []:
Is the information correct? [Y/n]
```

It is clear from the results that the adduser command did a significant amount of work for us. The command copied files from /etc/skel into our new user's home directory and set the user's home directory to /home/packt2 by default. The user account was also assigned the next available **User**

ID (UID) and **Group ID (GID)** of 1004. In point of fact, the adduser and useradd commands both copy files from the /etc/skel directory; however, the adduser command is somewhat more detailed in the tasks that it executes.

How to delete an account

When a user no longer needs access to a system, it is highly necessary to remove or disable their account because unmanaged accounts frequently become a security concern. This can be done by logging in to the system's administration panel and selecting the **Accounts** tab. We will make use of the userdel command in order to delete a user account. However, before you go and delete an account, there is one crucial question that you should consult with yourself about. Do you (or someone else) anticipate needing access to the user's files in some capacity? The vast majority of businesses have retention policies that outline what ought to take place with regard to a user's data in the event that they depart the organization. These files are occasionally duplicated and saved in an archive for long-term preservation. It's not uncommon for a manager, a coworker, or a new employee to require access to a previous user's files, possibly so that they can pick up where the previous user left off on a project. Before beginning to manage users, it is critical to have a solid understanding of this policy. If you do not already have a policy that describes the retention requirements for files when users depart the company, it is highly recommended that you collaborate with the management team to develop such a policy. The user's home directory's contents are not deleted when the userdel command is used since this behavior is not the default. In this section, we will delete packt2 from the system by executing the following command:

```
sudo userdel packt2
```

In this case, the home folder of the packt2 user remains.

If we want to remove the home folder once with the account, we need to use the -r parameter:

```
sudo userdel -r packt2
```

Before deleting users' accounts, remember to check whether the files in their home folders are needed. Once deleted, they cannot be recovered if there is no backup:

```
su - packt2
su: user packt2 does not exist
```

In conclusion, deleting a user account in Linux involves backing up data, terminating processes, removing the user from groups, deleting the home directory, updating system files, and performing a final cleanup. By following these steps, you can securely delete an account while managing the associated files and permissions effectively.

Understanding the/etc/sudoers file

In this section, let's see how to use the ordinary user account we created earlier to carry out user administration operations.

We must make a special permissions entry for packt in /etc/sudoers in order to allow it special access:

```
packt ALL=(ALL) ALL
```

Let's break down this line's syntax:

- First, we state to which user this rule applies (packt).

- All hosts that use the same /etc/sudoers file are covered by the rule if the first ALL is present. Since the same file is no longer shared among different machines, this term now refers to the current host.

- Next, (ALL) ALL informs us that any user may execute any command as the packt user. In terms of functionality, this is similar to (root) ALL.

It is important to manage permissions using groups as it makes life much easier. Imagine how simple it would be to just remove a user from a sudo group rather than removing the user from 100 different places.

Switching users

We are now prepared to begin using the packt account to carry out user administration duties. Use the su command to switch to that account to accomplish this. Notably, if you're using CentOS or a comparable operating system, you don't need to use the root account moving forward:

```
su -l packt
```

We are able to check the permissions for our newly formed packt account by using the sudo command. Let's create another account called packtdemo, shall we?

```
~ $sudo adduser packtdemo
Adding user `packtdemo' ...
Adding new group `packtdemo' (1005) ...
Adding new user `packtdemo' (1004) with group `packtdemo' ...
The home directory `/home/packtdemo' already exists.  Not copying from
`/etc/skel'.
New password:
Retype new password:
passwd: password updated successfully
Changing the user information for packtdemo
```

```
Enter the new value, or press ENTER for the default
Full Name []:
Room Number []:
Work Phone []:
Home Phone []:
Other []:
Is the information correct? [Y/n]
```

Changes to the user's home folder, default shell, and the ability to add a description to the user account can all be made with the usermod command.

Initially, the /etc/passwd file looks like this:

```
~ $grep packtdemo /etc/passwd
packtdemo:x:1004:1005:,,,:/home/packtdemo:/bin/bash
```

Let's try to add a description and change the shell:

```
~ $sudo usermod --comment "Demo account for packt" --shell /bin/sh
packtdemo
~ $grep packtdemo /etc/passwd
packtdemo:x:1004:1005:Demo account for packt:/home/packtdemo:/bin/sh
```

Changing to an alternate user account is frequently quite beneficial when working with support (especially while troubleshooting permissions). Take, for instance, the scenario in which a user complains to you that they are unable to access the files contained in a particular directory or that they are unable to execute a certain command. In such a scenario, you can try to duplicate the issue by logging in to the server, switching to the user account of the person having the issue, and then attempting to access the given files. In this way, you will not only be able to view their issue for yourself; you will also be able to test whether your solution resolves their problem before you report back to them.

You can switch to root user by running the sudo su or su – command, or just simply su.

su alone switches to another user while maintaining the current environment, while su – simulates a complete login environment for the target user, including setting up their home directory and environment variables and starting a new login shell. The choice between the two commands depends on the specific requirements or tasks you need to accomplish as the user switched to.

Managing account passwords

If you remember correctly, the passwd command enables us to alter the password for the user who is now logged in to the system. In addition, we are able to change the password for any user account on our system by running the passwd command while logged in as root and providing the username. However, that is only one of the capabilities of this command.

Locking/unlocking user accounts

The ability to lock and unlock a user account is one feature of the passwd command that we have not yet discussed. You can use this command to do either of these things. There are lots of different use cases where you need to accomplish something like this. For example, if a person is going to be gone for a lengthy period of time, you might want to lock their account so that it is inaccessible to other users during that time.

Use the -l option when you want to lock an account. For example, to lock the account for the packt user, we use the following command:

```
sudo passwd -l packt
```

Unlock it as follows:

```
sudo passwd -u packt
```

Setting password expiration

Next, we will discuss the process of actually carrying out the steps to set up password expiration.

To be more specific, the chage command is what enables us to do this. We may use chage to change the length of time for which a user's password is valid, and it also provides a more user-friendly alternative to reading the /etc/shadow file in order to view the current password expiration information. By giving a username and using the -l option of the chage command, we are able to view the pertinent information:

```
sudo chage -l packt
```

It is not necessary to run chage as root or with the sudo command. There is no need to raise your permission level to be able to view the expiration information for your own login. To access information using chage for any user account other than your own, however, you will need to utilize sudo:

```
~ $sudo chage -l packt
```

```
[voxsteel@centos8 ~]$ sudo chage -l packt
Last password change                                        : Feb 06, 2023
Password expires                                            : May 07, 2023
Password inactive                                           : never
Account expires                                             : never
Minimum number of days between password change              : 0
Maximum number of days between password change              : 90
Number of days of warning before password expires           : 7
[voxsteel@centos8 ~]$
```

Figure 7.1 – Display the password-related information and aging policies for a user account

In the output, we are able to view values such as the date that the password will expire, the maximum number of days that can pass before it needs to be changed, and so on. It's basically the same information that's saved in /etc/shadow, but it's formatted in a way that makes it much simpler to understand. If you would like to make any modifications to this information, the chage tool is once again your best option. The first illustration that I'll give you is a fairly typical one. You should absolutely require new users to reset their passwords the first time they log in after creating their user accounts for them. Unfortunately, not everyone will be interested in doing what has to be done. Using the chage command, you can require a user to change their password upon their first successful login to the system. This is done by changing their total number of days before password expiration to 0 in the following manner:

```
sudo chage -d 0 packt
```

And the results compared with the previous output look like this:

```
~ $sudo chage -l packt
```

```
[voxsteel@centos8 ~]$ sudo chage -l packt
Last password change                                      : Feb 06, 2023
Password expires                                          : never
Password inactive                                         : never
Account expires                                           : never
Minimum number of days between password change            : 0
Maximum number of days between password change            : 99999
Number of days of warning before password expires         : 7
[voxsteel@centos8 ~]$
```

Figure 7.2 – Display the password-related information and aging policies for a user account

The following instructions allow you to configure a user account so that it will demand a new password after a particular number of days have passed:

```
sudo chage -M 90 <username>
```

In the preceding code, I am configuring the user account to become invalid after 90 days and to demand a new password at that time. When the user logs in, they will be presented with a warning notice seven days before the password has to be changed. This is the message that will appear seven days before the password expires:

```
[voxsteel@centos8 ~]$ sudo chage -l packt
Last password change                                   : Feb 06, 2023
Password expires                                       : May 07, 2023
Password inactive                                      : never
Account expires                                        : never
Minimum number of days between password change         : 0
Maximum number of days between password change         : 90
Number of days of warning before password expires      : 7
[voxsteel@centos8 ~]$
```

Figure 7.3 – Display the password-related information and aging policies for a user account

It is a good practice to set a password expiration policy for security reasons.

Group management

Now that we know how to make new user accounts, manage existing user accounts, and switch between user accounts, we need to learn how to manage groups. Linux's implementation of the concept of groups is not all that dissimilar to that of other operating systems, and it essentially performs the same function. Controlling a user's access to the resources on your server can be done more effectively with the help of groups. You can grant access to users or deny access to users by simply adding them to or removing them from a group that has been assigned to a resource (such as a file or directory). This is made possible by assigning a group to the resource in question. The way that this is handled in Linux is such that each and every file and directory has both a user and a group that claims ownership of it. When using Linux, ownership is assigned on a one-to-one basis, meaning that each file or directory has just one user and just one group associated with it. You can verify this for yourself on a Linux system by listing the items included within a directory, as follows:

```
ls -l
```

```
[voxsteel@centos8 ~]$ ls -l
total 4
drwxr-xr-x. 3 voxsteel voxsteel  30 Jun 21  2022 Desktop
drwxr-xr-x. 2 voxsteel voxsteel   6 Dec 18  2021 Documents
drwxr-xr-x. 2 voxsteel voxsteel   6 Dec 18  2021 Downloads
drwxr-xr-x. 2 voxsteel voxsteel   6 Dec 18  2021 Music
drwxrwxr-x. 3 voxsteel voxsteel 141 Mar 20 14:18 packt
drwxr-xr-x. 2 voxsteel voxsteel   6 Dec 18  2021 Pictures
drwxr-xr-x. 2 voxsteel voxsteel   6 Dec 18  2021 Public
drwxr-xr-x. 2 voxsteel voxsteel   6 Dec 18  2021 Templates
-rw-r--r--. 1 voxsteel voxsteel 269 Jul 13  2022 test2.yml
drwxr-xr-x. 2 voxsteel voxsteel   6 Dec 18  2021 Videos
[voxsteel@centos8 ~]$
```

Figure 7.4 – Listing the contents of a folder

You only need to use the `cat` command to read the contents of the `/etc/group` file if you are interested in discovering which groups are currently active on your server. The `/etc/group` file, much like the `/etc/passwd` file that we went over before, stores information regarding the groups that have been created on your system. Feel free to take a look at this file, which is located on your system:

To create a group, use the `sudo addgroup <groupname>` command:

```
sudo addgroup packtgroup
Adding group `packtgroup' (GID 1006) ...
Done.
```

To view the content of the file, run the following command:

```
cat /etc/group
packtgroup:x:1006:
packt:x:1007:
packtdemo:x:1005:
packt2:x:1004:
```

The structure of each line is as follows:

```
<Group name>:<Group password>:<GID>:<Group members>
```

Examining the previous syntax, we can see the following:

- The name of the group is `<Group name>`
- Group passwords are not used if there is an x next to `<Group password>`
- `<Group members>` is a list, separated by commas, of users who are members of `<Group name>`

Use the `sudo delgroup <groupname>` command to remove a group.

Each line in the `/etc/group` file containing group information contains the group name and the user accounts connected to it:

```
packt:x:1007:
packtdemo:x:1005:
packt2:x:1004:
```

Now, we'll have a look at the `usermod` command, which is what actually allows you to associate users with groups and is the next topic on our agenda. The `usermod` command can be used for a variety of purposes and has a wide range of applications (adding a user to a group is just one of its abilities). In the event that we want to include a user (`packt`) in our `packtgroup` group, we would issue the following command:

```
sudo usermod -aG packtgroup packt
```

The -aG option is used to add a user to a specific group. The -a flag means *append*, which means that the user will be added to the group without removing them from any other groups they may already be a member of. The G flag specifies the group name.

If you wanted to modify the primary group that a user belonged to, you would use the -g option instead (note that it is written with a lowercase g rather than an uppercase G, as we did earlier):

```
sudo usermod -g <groupname> <username>
```

Groups are the easiest way to manage security permissions. Imagine removing one user from a group assigned to 100 resources rather than removing the user 100 times.

Permissions

In Linux, each file, directory, and other system object has a designated owner and group. This is the most fundamental aspect of system security that safeguards users from one another. Different sorts of access to read from, write to, or execute files can be given to owners, group members, and everyone else. In Linux, these are commonly referred to as file permissions.

The following commands are used to manage ownership and set permissions:

- Change file permissions with chmod
- Change the file owner with chown
- Change group ownership with chgrp
- Print the user and group IDs with id

Typically, the user who created a file is its owner, and the group attached to that owner is its primary group (at least initially). Let's create a testfile file in the /tmp directory as an example:

```
$echo "This is a test file" > testfile
$ls -l testfile
-rw-rw-r-- 1 packt packt 20 Feb  6 16:37 testfile
```

The first character of the output shows that testfile is a normal file (that is, not a directory or other type of system object). The next nine characters (split into three sets of three characters each) show the read (r), write (w), and execute (x) permissions for the system's owner, group owner, and other users.

The first three characters (rw-) show that the file's owner (user packt) can read and write to it. The next three characters show the same thing concerning the group owner. The last set of characters (r—) mean that other users can only read that file; they can't write to it or run it.

We'll use chmod to modify a file's permissions. A symbolic representation indicating to whom the new permissions will applied must come after this command:

- u means user
- g means group
- o means all other users
- a means all users

The types of permission are as follows:

- +r adds read permission
- -r removes read permission
- +w adds write permission
- -w removes write permission
- +x adds execute permission
- -x removes execute permission
- +rw adds read and write permissions
- +rwx adds read, write, and execute permissions

All of these permissions can be expressed numerically as follows:

Read – 4 : Write – 2 : Execute – 1

For example, chmod 754 testfile will be translated as rwx permissions on testfile for the owner, rx for the group, and only r for everyone else:

```
ls -la testfile
-rwxr-xr--  1 voxsteel voxsteel  0 26 Jun 14:27 testfile
```

To sum up, giving correct permissions is critical for security reasons and to avoid causing any unwanted damage. Following is an example of viewing the permissions:

```
[[voxsteel@centos8 packt]$ ls -la
total 16
drwxrwxr-x.  3 voxsteel voxsteel  109 Apr 21 14:08 .
drwx------. 18 voxsteel voxsteel 4096 Jan 18 14:07 ..
-rw-rw-r--.  1 voxsteel voxsteel   74 Nov 13 19:04 file1.txt
-rw-rw-r--.  1 voxsteel voxsteel   14 Oct  8  2022 file2.txt
-rw-rw-r--.  1 voxsteel voxsteel   56 Oct  8  2022 file3.txt
-rw-rw-r--.  1 voxsteel voxsteel    0 Oct  8  2022 file4.txt
-rw-rw-r--.  1 voxsteel voxsteel    0 Oct  8  2022 file5.txt
drwxrwxr-x.  2 voxsteel voxsteel   23 Oct  8  2022 testfolder
[voxsteel@centos8 packt]$
```

Figure 7.5 – Listing the contents of a folder and their permissions

To sum up, giving correct permissions is critical for security reasons to prevent any unwanted activity on your system.

Changing groups

The chgrp command will be discussed now in the context of making packtgroup the new owner of testfile. After the command, we specify the name of the group and the name of the file whose ownership is to be changed (in this case, testfile):

```
sudo chgrp packtgroup testfile
```

Let's check the ability of user packtdemo to write to this file now. A permission refused error should appear for the user. We can set the relevant permissions for the group to allow packtdemo to write to the file:

```
sudo chmod g+w testfile
```

Then use usermod once more to add the account to packtgroup, this time using the -aG combined option as follows:

```
sudo usermod -aG packtgroup packtdemo
```

The abbreviation for *append to group* is -aG.

Currently, packtgroup is referred to as a subsidiary or auxiliary group for user packtdemo. When packtdemo next logs in, the updated access permissions will be active.

We can use chown followed by the usernames and filenames, in that order, to make packtdemo the owner of testfile rather than just adding the user to the packtgroup group:

```
sudo chown packtdemo testfile
```

Keep in mind that the aforementioned command will prevent the `packt` user from accessing `testfile` because such an account is no longer the file's owner or a member of the `packtgroup` group.

Along with the typical `rwx` file permissions, `setuid`, `setgid`, and the sticky bit are three additional permissions that merit addressing. Let's look at them one by one.

Any user may run an executable file if the `setuid` bit is set on the file, utilizing the owner's permissions.

Any user may run an executable file when the `setgid` bit is set on the file, using the group's rights.

When misused, these specific rights present a security risk. For instance, if any user is permitted to run a command with superuser capabilities, that user will be able to access files that belong to root as well as to other users. It is simple to understand how this might quickly wreak havoc on a system: crucial files may be deleted, individual directories could be completely erased, and even hardware could end up acting erratically. All of this can be brought about by a single wicked or careless person. The `setuid` and `setgid` bits must therefore be used with extreme caution.

In `/usr/bin/passwd`, the `setuid` bit is required and is an acceptable use case. Although root owns this file, any user can change their own password by using it (but not that of other users).

When the sticky bit is set on a directory, no one other than the owner, the directory's owner, or root can delete files from that directory. This is commonly used to prevent a user from erasing the files of other users in a shared directory.

The `setuid` bit is set for `testfile` as follows:

```
sudo chmod u+s testfile
```

The `setgid` bit is set for `testfile` as follows:

```
sudo chmod g+s testfile
```

To create a directory named `packtdir` and set the sticky bit on it, use the following command:

```
sudo mkdir packtdir
sudo chmod +t packtdir
```

Going back to `/etc/sudoers`, we can also grant superuser access to every user in a group by using the `/etc/sudoers` file. For instance, the following command specifies that users who belong to `packtgroup` are permitted to run `updatedb` (or more specifically, `/usr/bin/updatedb`):

```
/usr/bin/updatedb %packtgroup ALL=(ALL)
```

The group name must be preceded by the `%` sign, which is the only distinction between group members and individual users. In this situation, command aliases are also applicable.

Simply type `sudo -l` in the command line and hit *Enter* to display the calling user's permitted commands in `/etc/sudoers`.

Using groups can save a lot of time. Imagine assigning one group permissions to run some commands, rather than 100 users one by one.

Summary

Managing users and permissions is something that will be required of you very frequently in sectors related to Linux administration, such as system administration and network security. This is something that will become embedded in your mental toolkit as new users join your organization while others depart. However, even if you are the only person who uses your servers, you will still need to manage permissions. This is due to the fact that processes are unable to run properly if they are denied access to the resources that they require in order to do their tasks.

In this chapter, we dove deep into the process of managing users, groups, and permissions and covered a lot of ground. We proceeded through the process of creating new users, removing existing users, assigning rights, and managing administrative access with the sudo command. Put these ideas into practice on your own server.

In the next chapter, we will talk about software installation and package repositories.

8

Software Installation and Package Repositories

Most Linux distributions only install the bare minimum set of software by default and assume that the user will install additional software later. These days, installing software on Linux is very easy, thanks to package repositories and high-level package managers that can search, download, and install packages over the network. However, it's important to understand how package files are organized internally and how package managers work since it helps administrators inspect packages, diagnose problems, and fix installation issues if they occur.

In this chapter, we'll cover the following topics:

- Software installation, packages, and dependencies
- Package files
- Package repositories and high-level package managers
- System upgrade

Software installation, packages, and dependencies

The definition of a software package is quite broad. In the early days of computing, when computers were exclusively used to solve mathematical problems, most programs were written completely from scratch to run on a specific computer, so there was no need for installation and thus no need for the concept of software packaging. For a long time afterward, software was still synonymous with executable files. Installing a software product that consists of a single executable file is trivial —just copy it to the target computer.

Such software certainly exists today. For example, the maintainers of **jq** (a popular tool for extracting data from JSON files) provide standalone, statically linked executables that combine the program and all libraries it needs into a monolithic file and can run on any Linux system, so any user can just download it from its website (`https://stedolan.github.io/jq/download/`) and start using it.

However, many software projects consist of multiple executable files and often require data files as well. A spell checker program, such as **Aspell**, requires dictionary files to work. In video games, the executable is often just a small part, often the smallest one compared to game assets such as models, textures, and sound files. To make the software product work as intended, all those files need to be distributed and installed together as a software package.

In early operating systems, the installation often just consisted of copying all files of a software product to a directory. That's the purpose of the /opt directory still found in many Linux systems. To install a hypothetical software package, Foo 1.0, an administrator may unpack its release archive (say, foo_1.0.tar.gz) to /opt/foo/ (usually, with tar --xf foo_1.0.tar.gz --directory /opt/foo/), and the directory structure inside /opt/foo will be completely defined by the application developers of Foo rather than by the operating system itself.

In modern operating systems, simply copying files is rarely enough because it will not properly integrate the newly installed software into the operating system. Even a package that provides a single console command needs to place it in a directory that is already in $PATH, or modify the $PATH environment variable accordingly. Software with a graphical user interface also needs to correctly register itself in the desktop environment application menus and optionally create desktop icons.

For this reason, operating systems started demanding a certain structure from software packages. In Microsoft Windows and many other systems, they would still usually come as custom executable programs that would unpack and copy the files and also execute scripts to register the package with the operating system. Developers of Linux software also practice this approach sometimes, and there are frameworks for creating executable installers, such as **makeself** (https://makeself.io/).

That approach gives software package authors complete control over the installation process, but it also has many disadvantages. First, uninstalling software that was installed that way isn't always reliable. If installer developers aren't careful to make it clean up all files created at installation time, users may be left with leftover files they will have to delete by hand.

Second, each package needs to implement its own update checking and installation mechanism. Before the widespread availability of broadband internet access, when software updates were usually distributed on physical media such as floppy disks or CDs, that wasn't a big concern; but these days, when most users have regular internet access, they benefit from online updates, and many software products such as web browsers must be kept up to date to keep them protected from security threats.

Third, few programs are written from scratch anymore and most make extensive use of third-party libraries, so they have dependencies. One way to make sure the library code is available for the program to call is to make it a part of the executable itself. That approach is known as static linking. It's foolproof in that there is no way to install such an executable incorrectly, but it leads to greatly increased memory and drive space consumption. For this reason, most programs are dynamically linked with external library files, but for them to work, they need to make sure that files of the correct versions of all required libraries exist on the system. If every software package is standalone and responsible for its own installation, the only way to do that is to bundle all libraries with the package (which is only marginally better than linking them statically, in terms of drive space usage) or require the user to find and install all libraries manually (which is very inconvenient for the user).

With proprietary software that isn't redistributable and usually isn't available in source code form, bundling all dependencies is usually the only option since executables need to be recompiled for use with different library versions. Different Linux distributions may include different versions of libraries due to different release cycles and software inclusion policies (some distributions are focused on stability and will only include older but proven versions, while others may choose to include the latest libraries even if they aren't well tested).

However, with open source software that is available in source code form and can be modified and redistributed, there are many more possibilities to reuse library files and create a cohesive system. To make that possible, distribution maintainers developed modern package managers.

Package managers

A package manager is a program that is responsible for the installation, upgrade, and removal of software packages. In the older approach where software developers are responsible for the installation procedure, they are free to choose whether to distribute their software as an executable that unpacks and installs files or as an archive that the user must unpack manually, and the choice of archive and compression algorithms is also on the developers.

By contrast, package managers usually use a very precise definition of a software package. To make a software project installable with a package manager, its maintainer must create a package file that follows a set of guidelines for its internal structure. Apart from the files needed for a software product to work, package files also contain metadata in a specific format. Metadata files contain information about the package, such as its name, version, license, and lists of other packages that it depends on.

Package managers limit what software product authors can do at installation time. For example, there is no way to let the user specify a custom installation directory or choose not to install certain files.

But since they completely control the installation process and every file inside the package is accounted for, they can install and uninstall software reliably and ensure that there are no files left after uninstallation.

An even bigger advantage is that package managers can automatically track package dependencies and either prevent the user from installing a package until all dependencies are installed or automatically install those dependencies.

Different Linux distributions use different package managers. Some of them use two different utilities for managing software installations and upgrades: a low-level package manager that is responsible for working with package files and a high-level tool for automatically downloading packages over the network and managing upgrades.

Historically, that was the dominant approach: before the widespread availability of broadband Internet access, automatic download wasn't feasible, so the first package managers developed in the 1990s were only developed to work with local package files. The two most popular projects in this category are **rpm** (developed by Red Hat) and **dpkg** (developed by the Debian project). For the automated installation of packages over the network and automated system upgrades, they need to be combined with high-level tools such as YUM, DNF, Zypper, or **Advanced Packaging Tool (APT)**.

Some Linux distributions developed in the 2000s and later use package managers that combine both functions and can work with local package files and remote sources alike.

We can represent current situations with low-level and high-level package managers in different distributions as in the following table:

Distribution	High-level package manager	Low-level package manager
Debian	APT	dpkg
Ubuntu		
Linux Mint		
Fedora	DNF	rpm
Red Hat Enterprise Linux	DNF, YUM	
openSUSE, SUSE Linux Enterprise Server	Zypper	
Mageia	DNF, urpmi	
Arch Linux	pacman	
NixOS	Nix	
Guix	Guix	

Table 8.1 – Package managers used in different Linux distributions

We will focus on the most popular low-level package managers: rpm and dpkg. Neither of them is inherently superior to the other, but their package file format and package management tool interface design choices are quite different. We will examine both to compare and contrast them.

Package files

Conventionally, packages come as files in a special format that includes both files from the software project that must be installed in the system and metadata for the package manager, such as file checksums, lists of dependencies, and system requirements for the package (such as CPU architecture). We will look inside .rpm and .deb package files and compare their implementation details.

Inspecting package files

First, we will examine package files to see what's inside them and learn how to examine and unpack them.

Please note that normally, you will not need to manually download and unpack package files! We are doing it only for educational purposes.

Inspecting a Debian package

We will use the GNU `hello` package for our experiments. GNU hello is a demo program that simply prints `hello world`—its real purpose is to serve as an example of development and packaging practices and show new developers how to write build scripts, implement internationalization, and so on.

You can find the GNU `hello` package from the latest unstable Debian version at `https://packages.debian.org/sid/hello`.

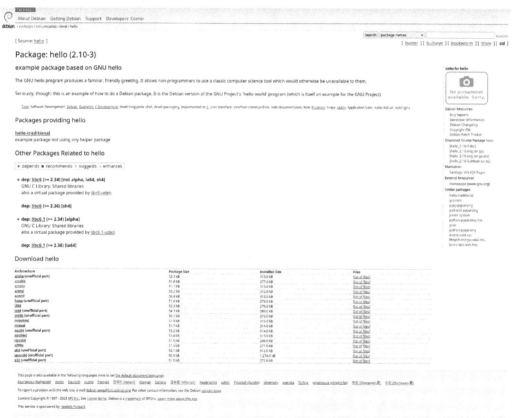

Figure 8.1 – The hello package information page on the Debian package repository website

Follow an architecture-specific download link such as https://packages.debian.org/sid/amd64/hello/download and download the package file from any mirror from the list shown in the following screenshot:

Figure 8.2 – The download page for the hello package for Debian Sid on x86_64 machines

At the time of writing, the latest version is 2.10, so I used this link, but it's not guaranteed to work for future versions:

```
$ wget http://ftp.us.debian.org/debian/pool/main/h/hello/hello_2.10-2_
amd64.deb
```

Using the dpkg --info command, we can view information about the package we have just downloaded:

```
$ dpkg --info ./hello_2.10-2_amd64.deb
new Debian package, version 2.0.
size 56132 bytes: control archive=1868 bytes.
Package: hello
Version: 2.10-2
Architecture: amd64
Maintainer: Santiago Vila <sanvila@debian.org>
Installed-Size: 280
Depends: libc6 (>= 2.14)
Conflicts: hello-traditional
Breaks: hello-debhelper (<< 2.9)
Replaces: hello-debhelper (<< 2.9), hello-traditional
Section: devel
Priority: optional
Homepage: http://www.gnu.org/software/hello/
Description: example package based on GNU hello
The GNU hello program produces a familiar, friendly greeting. It
allows non-programmers to use a classic computer science tool which
would otherwise be unavailable to them.
 .
Seriously, though: this is an example of how to do a Debian package.
It is the Debian version of the GNU Project's `hello world' program
(which is itself an example for the GNU Project).
```

Where does dpkg get that information from? Notice the control archive=1868 bytes part of the output. A Debian package file is an ar archive that consists of two compressed tar archives glued together. We could certainly extract them using the ar utility, or even with dd (thanks to the simplicity of its format and the fact that dpkg tells us each archive length in bytes), and unpack them by hand, but luckily, dpkg has built-in functionality for that. Using dpkg --control (dpkg -e), we can extract the control archive—the part that contains package metadata:

```
$ dpkg --control ./hello_2.10-2_amd64.deb
```

If we don't specify a custom output directory, dpkg will unpack it into a subdirectory named DEBIAN. It will contain two files: control and md5sums. The control file is where dpkg --info (dpkg -l) took those fields and their values from:

```
$ cat DEBIAN/control
Package: hello
Version: 2.10-2
Architecture: amd64
Maintainer: Santiago Vila <sanvila@debian.org>
Installed-Size: 280
Depends: libc6 (>= 2.14)
Conflicts: hello-traditional
Breaks: hello-debhelper (<< 2.9)
Replaces: hello-debhelper (<< 2.9), hello-traditional
Section: devel
Priority: optional
Homepage: http://www.gnu.org/software/hello/
Description: example package based on GNU hello
The GNU hello program produces a familiar, friendly greeting. It
allows non-programmers to use a classic computer science tool which
would otherwise be unavailable to them.
.
Seriously, though: this is an example of how to do a Debian package.
It is the Debian version of the GNU Project's `hello world' program
(which is itself an example for the GNU Project).
```

The md5sums file contains hash sums of all files inside the package:

```
$ cat DEBIAN/md5sums
6dc2cf418e6130569e0d3cfd2eae0b2e usr/bin/hello
9dbc678a728a0936b503c0aef4507f5d usr/share/doc/hello/NEWS.gz
1565f7f8f20ee557191040f63c1726ee
usr/share/doc/hello/changelog.Debian.gz
31aa50363c73b22626bd4e94faf90da2 usr/share/doc/hello/changelog.gz
bf4bad78d5cf6787c6512b69f29be7fa usr/share/doc/hello/copyright
c5162d14d046d9808bf12adac2885473 usr/share/info/hello.info.gz
0b430a48c9a900421c8d2a48e864a0a5 usr/share/locale/bg/LC_MESSAGES/
hello.mo
0ac1eda691cf5773c29fb65ad5164228 usr/share/locale/ca/LC_MESSAGES/
hello.mo
...
usr/share/locale/zh_TW/LC_MESSAGES/hello.mo
29b51e7fcc9c18e989a69e7870af6ba2 usr/share/man/man1/hello.1.gz
```

The MD5 hash sum algorithm is no longer cryptographically secure and must not be used to protect files and messages from malicious modification. However, in Debian packages, it is used only for protection against accidental file corruption, so it's not a security issue. Protection against malicious modification is provided by GnuPG digital signatures, which are checked by the high-level tool—APT.

If you only want to list files inside the data archive, you can do it with dpkg --contents (dpkg -c):

```
$ dpkg --contents ./hello_2.10-2_amd64.deb
drwxr-xr-x root/root  0 2019-05-13 14:06 ./
drwxr-xr-x root/root 0 2019-05-13 14:06 ./usr/
drwxr-xr-x root/root 0 2019-05-13 14:06 ./usr/bin/
-rwxr-xr-x root/root31360 2019-05-13 14:06
./usr/bin/hello
drwxr-xr-x root/root 0 2019-05-13 14:06
./usr/share/
drwxr-xr-x root/root 0 2019-05-13 14:06
./usr/share/doc/
drwxr-xr-x root/root 0 2019-05-13 14:06 ./usr/share/doc/hello/
-rw-r--r-- root/root 1868 2014-11-16 06:51 ./usr/share/doc/hello/NEWS.
gz
-rw-r--r-- root/root 4522 2019-05-13 14:06 ./usr/share/doc/hello/
changelog.Debian.gz
-rw-r--r-- root/root 4493 2014-11-16 07:00 ./usr/share/doc/hello/
changelog.gz
-rw-r--r-- root/root 2264 2019-05-13 13:00 ./usr/share/doc/hello/
copyright
drwxr-xr-x root/root 0 2019-05-13 14:06 ./usr/share/info/
-rw-r--r-- root/root  11596 2019-05-13 14:06
./usr/share/info/hello.info.gz
drwxr-xr-x root/root 0 2019-05-13 14:06 ./usr/share/locale/
...
```

It's also possible to unpack the data archive part of a package file with dpkg --extract (dpkg -x), but in that case, you need to specify where to unpack it. To unpack into the current directory, we can use the dot shortcut:

```
$ dpkg --extract ./hello_2.10-2_amd64.deb .
```

Now, let's move on to inspecting an RPM package file and comparing it with what we have seen in a Debian package.

Inspecting an RPM package

We'll use Fedora as an example of a distribution that uses RPM packages. The place to search Fedora repositories on the web is `https://packages.fedoraproject.org/`. Enter the name of the package (`hello`, in our case) in the search field and you'll see the package information page.

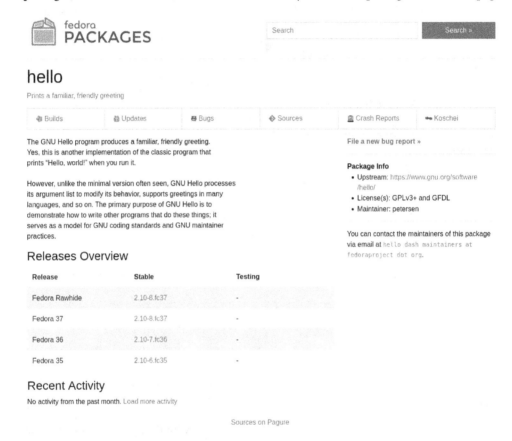

Figure 8.3 – The hello package information page on the Fedora package repository website

Fedora uses Rawhide as a code name for the latest unstable version. At the time of writing, the version in development is 37. To find a package download link, go to the **Builds** tab, then find a link to the latest build there. We can download that package for inspection using the following command:

```
$ wget https://kojipkgs.fedoraproject.org//packages/hello/2.10/8.fc37/
x86_64/hello-2.10-8.fc37.x86_64.rpm
```

> **Note**
>
> We have provided a sample direct link to the package from Fedora 37, but since all URLs and package versions are subject to change, the following command is not guaranteed to remain working forever—if you want to download the package, visit the `packages.fedoraproject.org` website instead.

All RPM queries are available through the `--query` option. A good thing about RPM is that all query options can be used to inspect either package files or already installed packages, depending on the argument. If you give it just a package name, it will look for an installed package, but if you specify a package file path, it will inspect that file instead.

For example, with `rpm --query --info`, we can read package metadata. Alternatively, that command can be shortened to `rpm -qi`:

```
$ rpm --query --info ./hello-2.10-8.fc37.x86_64.rpm
Name        : hello
Version     : 2.10
Release     : 8.fc37
Architecture: x86_64
Install Date: (not installed)
Group       : Unspecified
Size        : 183537
License     : GPLv3+ and GFDL
Signature   : (none)
Source RPM  : hello-2.10-8.fc37.src.rpm
Build Date  : Thu 21 Jul 2022 03:15:21 PM IST
Build Host  : buildvm-x86-27.iad2.fedoraproject.org
Packager    : Fedora Project
Vendor      : Fedora Project
URL         : https://www.gnu.org/software/hello/
Bug URL     : https://bugz.fedoraproject.org/hello
Summary     : Prints a familiar, friendly greeting
Description :
The GNU Hello program produces a familiar, friendly greeting.
Yes, this is another implementation of the classic program that prints
"Hello, world!" when you run it.

However, unlike the minimal version often seen, GNU Hello processes
its argument list to modify its behavior, supports greetings in
many languages, and so on. The primary purpose of GNU Hello is
to demonstrate how to write other programs that do these things;
it serves as a model for GNU coding standards and GNU maintainer
practices.
```

Using `rpm --query --list` (or `rpm -ql` for short), we can get a list of files inside the package:

```
$ rpm --query --list ./hello-2.10-8.fc37.x86_64.rpm
/usr/bin/hello
/usr/lib/.build-id
/usr/lib/.build-id/d2
/usr/lib/.build-id/d2/847d989fd9b360c77ac8afbbb475415401fcab
/usr/share/info/hello.info.gz
/usr/share/licenses/hello
/usr/share/licenses/hello/COPYING
/usr/share/locale/bg/LC_MESSAGES/hello.mo
/usr/share/locale/ca/LC_MESSAGES/hello.mo
...
/usr/share/locale/zh_CN/LC_MESSAGES/hello.mo
/usr/share/locale/zh_TW/LC_MESSAGES/hello.mo
/usr/share/man/man1/hello.1.gz
```

Compare and contrast this with the output of `dpkg --contents`, which listed MD5 sums for every file. Unlike Debian packages, RPM packages do not contain hash sums for every file, but rather a single hash sum of the package archive.

The `rpm` command also does not provide options for unpacking RPM packages. Instead, it provides two utilities for extracting its data part: `rpm2cpio` and `rpm2archive`.

Unlike Debian packages, RPM packages do not store metadata as files inside an archive, but rather use a custom RPM file header to store it. The archive that follows that header only contains files that must be installed. Moreover, while `dpkg` uses a familiar `tar` archive format to pack multiple files into one, RPM uses a much less common CPIO format.

The `cpio` command is difficult to use. In particular, it needs the user to enter the path to every file that must be included in the archive, so it's impractical to use by hand and can only be reasonably used in conjunction with another tool, such as `find`. For this reason, `tar` is far more popular because the TAR archiving tool can easily pack entire directories in one command, such as `tar cvf file. tar /path/to/directory`. However, the CPIO archive file format is simpler to implement, doesn't vary between implementations from different vendors, and has better support for special files, such as links and device files, which is why some projects have chosen to use it internally. Among those projects are the Linux kernel, which uses it for the initial RAM disk, and RPM, which uses it for package files.

Luckily, the `rpm2archive` utility can convert an RPM package into a compressed `tar` archive, so you don't need to learn how to use `cpio` just to extract files from RPM packages:

```
$ rpm2archive hello-2.10-8.fc37.x86_64.rpm
$ tar xvfz ./hello-2.10-8.fc37.x86_64.rpm.tgz
./usr/bin/hello
```

```
./usr/lib/.build-id/
./usr/lib/.build-id/d2/
./usr/lib/.build-id/d2/847d989fd9b360c77ac8afbbb475415401fcab
./usr/share/info/hello.info.gz
./usr/share/licenses/hello/
./usr/share/licenses/hello/COPYING
./usr/share/locale/bg/LC_MESSAGES/hello.mo
./usr/share/locale/ca/LC_MESSAGES/hello.mo
...
```

Now let's move on to inspecting packages installed in the system.

Inspecting installed packages

Both software project files and the metadata that we have seen inside package files are retained in the system when the package is installed. Let's learn how to access information about installed packages and compare it with what we have seen inside package files.

Listing all installed packages

Both dpkg and rpm provide an option to list all packages installed on the system. Since those lists are going to be rather long even for small installations, you may want to use them with a pager command such as less or a filter such as head, tail, or grep.

For Debian-based systems, the command to list all installed packages is dpkg --list, or dpkg -l for short. The list is sorted alphabetically by default:

```
$ dpkg -l | head
Desired=Unknown/Install/Remove/Purge/Hold
| Status=Not/Inst/Conf-files/Unpacked/halF-conf/Half-inst/trig-aWait/
Trig-pend
|/ Err?=(none)/Reinst-required (Status,Err: uppercase=bad)
||/ Name                          Version
Architecture Description
ii  adduser                       3.118
all          add and remove users and groups
ii  adwaita-icon-theme            3.30.1-1
all          default icon theme of GNOME
ii  ansible                       2.9.27-1ppa~trusty
all          Ansible IT Automation
ii  apparmor                      2.13.2-10
amd64        user-space parser utility for AppArmor
ii  apt                           1.8.2.3
amd64        commandline package manager
```

For RPM-based distributions, that command is `rpm --query --all`, or `rpm -qa`. Note that RPM does not sort that output alphabetically by default:

```
$ rpm --query --all | head
shared-mime-info-2.1-3.fc35.x86_64
xorg-x11-drv-vmware-13.2.1-16.fc35.x86_64
xorg-x11-drv-qxl-0.1.5-20.fc35.x86_64
irqbalance-1.7.0-8.fc35.x86_64
ipw2200-firmware-3.1-22.fc35.noarch
ipw2100-firmware-1.3-29.fc35.noarch
gpg-pubkey-9867c58f-601c49ca
gpg-pubkey-d651ff2e-5dadbbc1
gpg-pubkey-7fac5991-4615767f
gpg-pubkey-d38b4796-570c8cd3
```

If you simply want to check whether a package is installed, RPM provides an option to check just that, without outputting any other information, that is, `rpm --query` (`rpm -q`):

```
$ rpm -q rpm
rpm-4.17.1-3.fc36.x86_64
$ rpm -q no-such-package
package no-such-package is not installed
```

The equivalent for Debian-based systems is `dpkg --status` (`dpkg -s`), which also prints information about a package if it's installed:

```
$ dpkg -s dpkg
Package: dpkg
Essential: yes
Status: install ok installed
Priority: required
Section: admin
Installed-Size: 6693
Maintainer: Dpkg Developers <debian-dpkg@lists.debian.org>
Architecture: amd64
Multi-Arch: foreign
Version: 1.19.7
Depends: tar (>= 1.28-1)
Pre-Depends: libbz2-1.0, libc6 (>= 2.15), liblzma5 (>= 5.2.2),
libselinux1 (>= 2.3), zlib1g (>= 1:1.1.4)
Suggests: apt, debsig-verify
Breaks: acidbase (<= 1.4.5-4), amule (<< 2.3.1+git1a369e47-3), beep
(<< 1.3-4), im (<< 1:151-4), libapt-pkg5.0 (<< 1.7~b), libdpkg-perl
(<< 1.18.11), lsb-base (<< 10.2019031300), netselect (<< 0.3.ds1-27),
pconsole (<< 1.0-12), phpgacl (<< 3.3.7-7.3), pure-ftpd (<< 1.0.43-1),
systemtap (<< 2.8-1), terminatorx (<< 4.0.1-1), xvt (<= 2.1-20.1)
```

```
Conffiles:
 /etc/alternatives/README 7be88b21f7e386c8d5a8790c2461c92b
 /etc/cron.daily/dpkg f20e10c12fb47903b8ec9d282491f4be
 /etc/dpkg/dpkg.cfg f4413ffb515f8f753624ae3bb365b81b
 /etc/logrotate.d/alternatives 5fe0af6ce1505fefdc158d9e5dbf6286
 /etc/logrotate.d/dpkg 9e25c8505966b5829785f34a548ae11f
Description: Debian package management system
This package provides the low-level infrastructure for handling the
installation and removal of Debian software packages.
 .
For Debian package development tools, install dpkg-dev.
Homepage: https://wiki.debian.org/Teams/Dpkg
```

If a package is not installed, that command prints an error:

```
$ dpkg -s no-such-package
dpkg-query: package 'no-such-package' is not installed and no
information is available
Use dpkg --info (= dpkg-deb --info) to examine archive files.
```

If you want to retrieve information about an installed package in an RPM-based system, you can use the same rpm -qi command as we used for inspecting a package file; just give it the name of the package rather than a file path:

```
$ rpm --query --info rpm
Name         : rpm
Version      : 4.17.1
Release      : 3.fc36
Architecture: x86_64
Install Date: Tue 09 Aug 2022 03:05:25 PM IST
Group        : Unspecified
Size         : 2945888
License      : GPLv2+
Signature    : RSA/SHA256, Tue 02 Aug 2022 02:11:12 PM IST,
Key ID 999f7cbf38ab71f4
Source RPM   : rpm-4.17.1-3.fc36.src.rpm
Build Date   : Tue 02 Aug 2022 01:31:56 PM IST
Build Host   : buildhw-x86-11.iad2.fedoraproject.org
Packager     : Fedora Project
Vendor       : Fedora Project
URL          : http://www.rpm.org/
Bug URL      : https://bugz.fedoraproject.org/rpm
Summary      : The RPM package management system
Description :
The RPM Package Manager (RPM) is a powerful command line driven
```

```
package management system capable of installing, uninstalling,
verifying, querying, and updating software packages. Each software
package consists of an archive of files along with information about
the package like its version, a description, etc.
```

If you need to find out what package a file comes from (or whether it belongs to any package at all), there are commands for that task as well. For RPM, it's rpm --query --file (rpm -qf):

```
$ rpm --query --file /usr/bin/rpm
rpm-4.17.1-3.fc36.x86_64

$ touch /tmp/test-file

$ rpm --query --file /tmp/test-file
file /tmp/test-file is not owned by any package
```

For dpkg, it's dpkg --search (dpkg -S):

```
$ dpkg -S /usr/bin/dpkg
dpkg: /usr/bin/dpkg
$ touch /tmp/test-file

$ dpkg -S /tmp/test-file
dpkg-query: no path found matching pattern /tmp/test-file
```

With RPM, it's also easy to list all files that belong to a certain package, and the command is the same as we have already used to list the contents of a package file:

```
$ rpm --query --list rpm
/etc/rpm
/usr/bin/rpm
/usr/bin/rpm2archive
/usr/bin/rpm2cpio
/usr/bin/rpmdb
/usr/bin/rpmkeys
/usr/bin/rpmquery
/usr/bin/rpmverify
...
```

For dpkg, the equivalent is dpkg --listfiles (dpkg -L). However, dpkg also lists all directories where files or subdirectories from that package can be found, even top-level directories such as / etc, while RPM only shows files and directories that were created as a result of installing a package:

```
$ dpkg -L dpkg | head
/.
/etc
```

```
/etc/alternatives
/etc/alternatives/README
/etc/cron.daily
/etc/cron.daily/dpkg
/etc/dpkg
/etc/dpkg/dpkg.cfg
/etc/dpkg/dpkg.cfg.d
/etc/logrotate.d
```

However, installing packages and removing them is a much more frequent task than inspecting installed ones—let's learn how to do it.

Installing and removing package files

Even the man page of dpkg warns that unpacking a package to the root directory is not the correct way to install it. There are multiple reasons why the installation of a package is not as simple as just copying the files from it to the right places.

Apart from copying files, package managers record the package and its files in an internal database— that is why commands such as dpkg --listfiles and rpm --query --files can work, and why package managers can reliably delete packages without leaving any unused files behind.

However, package managers also include safeguards against trying to install a package that is guaranteed not to work on the system. For example, this is what will happen if you try to install a package built for ARM64 on an x86_64 (Intel or AMD64) machine:

```
$ wget http://ftp.us.debian.org/debian/pool/main/h/hello/hello_2.10-2_
arm64.deb
$ sudo dpkg --install ./hello_2.10-2_arm64.deb
dpkg: error processing archive ./hello_2.10-2_arm64.deb (--install):
package architecture (arm64) does not match system (amd64)
Errors were encountered while processing:
./hello_2.10-2_arm64.deb
```

They will also protect the system from attempts to delete packages that should not be deleted, or at least warn the user about the consequences. Since they track all dependencies between packages, they can force the removal of packages that will become broken if a package they depend on is removed, but by default, they will refuse to delete any package if other packages depend on it.

In Debian-based Linux distributions, there's a concept of essential packages that are protected from removal attempts. For example, you will get an error if you try to delete Bash since it's the default system shell:

```
$ sudo dpkg --remove bash
dpkg: error processing package bash (--remove):
```

```
   this is an essential package; it should not be removed
Errors were encountered while processing:
 bash
```

If a package is not essential, you will get a list of dependent packages that prevent its removal:

```
$ sudo dpkg --remove gcc
dpkg: dependency problems prevent removal of gcc:
musl-tools depends on gcc.
g++ depends on gcc (= 4:8.3.0-1).
dkms depends on gcc.
build-essential depends on gcc (>= 4:8.3).
```

RPM has a similar functionality and will not allow the user to install packages with unsatisfied (or unsatisfiable, in case of a different architecture) dependencies or remove essential packages.

This is what an attempt to install a package for a different architecture may look like:

```
$ sudo rpm --install hello-2.10-8.fc37.aarch64.rpm
error: Failed dependencies:
ld-linux-aarch64.so.1()(64bit) is needed by hello-2.10-8.fc37.aarch64
ld-linux-aarch64.so.1(GLIBC_2.17)(64bit) is needed by hello-2.10-8.
fc37.aarch64
```

And this is what you will get if you try to remove rpm itself:

```
$ sudo rpm --erase rpm
error: Failed dependencies:
rpm is needed by (installed) policycoreutils-3.3-4.fc36.x86_64
rpm is needed by (installed) cmake-rpm-macros-3.22.2-1.fc36.noarch
rpm is needed by (installed) kde-filesystem-4-67.fc36.x86_64
rpm is needed by (installed) color-filesystem-1-28.fc36.noarch
rpm is needed by (installed) efi-srpm-macros-5-5.fc36.noarch
rpm is needed by (installed) lua-srpm-macros-1-6.fc36.noarch
rpm > 4.15.90-0 is needed by (installed) python3-rpm-generators-12-15.
fc36.noarch
rpm = 4.17.1-3.fc36 is needed by (installed) rpm-libs-4.17.1-3.fc36.
x86_64
rpm = 4.17.1-3.fc36 is needed by (installed) rpm-build-4.17.1-3.fc36.
x86_64
rpm is needed by (installed) rpmautospec-rpm-macros-0.3.0-1.fc36.
noarch
rpm is needed by (installed) python3-rpmautospec-0.3.0-1.fc36.noarch
rpm >= 4.9.0 is needed by (installed) createrepo_c-0.20.1-1.fc36.
x86_64
rpm >= 4.15 is needed by (installed) fedora-gnat-project-
```

```
common-3.15-4.fc36.noarch
rpm >= 4.11.0 is needed by (installed) redhat-rpm-config-222-1.fc36.
noarch
```

However, in modern systems, both the installation and removal of packages with rpm and dpkg are an exception rather than a rule. While they prevent the installation of packages that have missing dependencies, they do not automate the installation of those dependencies, so trying to install a package with a lot of dependencies is a very tedious task. If you were to upgrade your system, you'd also have to upgrade all packages and their dependencies one by one. That's the main reason why package repositories and high-level package managers were invented.

Package repositories and high-level package managers

Online collections of packages for Linux distributions have existed for almost as long as the distributions themselves. They saved users' time searching for compiled packages or building software from source, but if a package had dependencies, the user would still need to download them all one by one.

The next step for distribution maintainers was to create a format for machine-readable metadata from the entire package collection and a tool that would automate that process. Since every package contains information about its dependencies, in the simplest case, you just need to download them all.

In reality, it's more complicated. Packages may conflict (for example, because they provide an executable with the same name) and there must be a safeguard against attempts to install conflicting packages. If a user tries to install a package from outside the repository, the repository may not have the right versions of its dependencies. Modern high-level package managers check for these and many other possible issues, so most of the time the user can just say "*I want a certain package installed*" and the tool will do everything that needs to be done to install it.

Package repositories

A package repository in the modern sense is a collection of packages that comes with a machine-readable index. The repository index contains information about every package. Essentially, an index aggregates the package metadata files we have seen when inspecting packages with dpkg --info and rpm --query --info.

Websites with information about distribution packages such as https://www.debian.org/distrib/packages and https://packages.fedoraproject.org/ are in fact generated from their package repository indices. When we used them to look up and download packages, we essentially did by hand what high-level package managers can do automatically.

Every distribution has its official repositories, and community- or vendor-supported repositories exist for popular distributions.

Debian-based distribution repositories

Repository configuration in Debian-based distributions can be found in the /etc/apt/ directory. The main file is /etc/apt/sources.list but there may be additional files in /etc/apt/sources.list.d.

This is an example of a sources.list file from a Debian 10 (Buster) system. Notice that apart from remote repositories, it's also possible to specify paths to repositories on local media such as optical drives. If you install a Debian system from a CD or DVD, the installer will add that as a repository entry so that you can install additional packages from that disk later on:

```
$ cat /etc/apt/sources.list
#deb cdrom:[Debian GNU/Linux 10.5.0 _Buster_ - Official amd64 NETINST
20200801-11:34]/ buster main
#deb cdrom:[Debian GNU/Linux 10.5.0 _Buster_ - Official amd64 NETINST
20200801-11:34]/ buster main
deb http://deb.debian.org/debian/ buster main
deb-src http://deb.debian.org/debian/ buster main
deb http://security.debian.org/debian-security buster/updates main
deb-src http://security.debian.org/debian-security buster/updates main
# buster-updates, previously known as 'volatile'
deb http://deb.debian.org/debian/ buster-updates main
deb-src http://deb.debian.org/debian/ buster-updates main
```

Debian also uses different repository components that you can enable or disable. In this example, only the main repository is enabled. The main repository includes essential packages. Apart from them, there are contrib and non-free repositories. The contrib repository includes a broad selection of additional packages that aren't as actively maintained as those from main. The non-free repository contains packages whose licenses don't qualify as free software—they are redistributable, but may have restrictions on modifications or distribution of modified versions, or come without source code.

For better or worse, many firmware files required for devices to work are under non-free licenses, so in practice, you may always use main contrib non-free rather than just main in Debian repository configuration lines.

Notice that the distribution version (buster, in this case) is explicitly hardcoded in the configuration file and you will need to change it yourself to upgrade to a new distribution version.

The repository URL is also set explicitly, usually to the mirror site you chose at installation time.

RPM package repositories

Fedora and Red Hat Enterprise Linux (and its derivatives such as CentOS and Rocky Linux) keep repository files in /etc/yum.repos.d/. Most often, there is one file per repository. This is an example of the base Fedora repository file:

```
$ cat /etc/yum.repos.d/fedora.repo
[fedora]
name=Fedora $releasever - $basearch
#baseurl=http://download.example/pub/fedora/linux/
releases/$releasever/Everything/$basearch/os/
metalink=https://mirrors.fedoraproject.org/metalink?repo=fedora-
$releasever&arch=$basearch
enabled=1
countme=1
metadata_expire=7d
repo_gpgcheck=0
type=rpm
gpgcheck=1
gpgkey=file:///etc/pki/rpm-gpg/RPM-GPG-KEY-fedora-$releasever-
$basearch
skip_if_unavailable=False
[fedora-debuginfo]
name=Fedora $releasever - $basearch - Debug
#baseurl=http://download.example/pub/fedora/linux/
releases/$releasever/Everything/$basearch/debug/tree/
metalink=https://mirrors.fedoraproject.org/metalink?repo=fedora-debug-
$releasever&arch=$basearch
enabled=0
metadata_expire=7d
repo_gpgcheck=0
type=rpm
gpgcheck=1
gpgkey=file:///etc/pki/rpm-gpg/RPM-GPG-KEY-fedora-$releasever-
$basearch
skip_if_unavailable=False
[fedora-source]
name=Fedora $releasever - Source
#baseurl=http://download.example/pub/fedora/linux/
releases/$releasever/Everything/source/tree/
metalink=https://mirrors.fedoraproject.org/metalink?repo=fedora-
source-$releasever&arch=$basearch
enabled=0
metadata_expire=7d
repo_gpgcheck=0
type=rpm
```

```
gpgcheck=1
gpgkey=file:///etc/pki/rpm-gpg/RPM-GPG-KEY-fedora-$releasever-
$basearch
skip_if_unavailable=False
```

Notice that there is no explicit reference to the Fedora version anywhere; instead, there are placeholder variables such as $releasever and $basearch. The high-level management tools automatically substitute those variables with data from the installed system.

There is also an option to make the system use mirror lists instead of single mirrors for reliability and load balancing. You can specify either a specific mirror in the baseurl option or a link to the Metalink protocol to automatically get a mirror list instead.

High-level package managers

As we already discussed, both rpm and dpkg were developed at a time when automatically downloading packages over a network was not feasible for most users because most computers only had a slow and intermittent connection to the Internet over a dial-up modem. By the end of the 90s, Internet connections had become significantly faster, so online package repositories became a logical possibility.

The Debian project developed its high-level package manager named APT in 1998 and most Debian-based distributions have been using it since then.

RPM-based distributions developed multiple high-level package managers independently. One reason for that is that many of the RPM-based distributions were created independently in the 90s, while most Debian-based ones are forks of Debian itself that were created after the introduction of APT.

Linux distributions maintained by Red Hat Inc. itself went through multiple high-level package managers, starting from up2date, which was mainly meant for their paying customers. That tool was later replaced by YUM, which came from a now-defunct distribution named Yellow Dog Linux. Its name originally meant Yellow dog Updater, Modified. Later, it was replaced by DNF—since the mid-2010s in Fedora and since the eighth release of Red Hat Enterprise Linux in 2019. Older but still supported versions of Red Hat Enterprise Linux and its derivatives continue to use YUM. Luckily for users, while YUM and DNF mostly differ in their internal implementation, their user interface is almost the same—except for the new features of DNF, of course. The name DNF doesn't officially stand for anything, but originally came from **DaNdiFied YUM**.

The design choices of APT and YUM/DNF are very different, and you need to be aware of those differences when you switch between them. Let's learn how to use them to install packages.

Installing and removing packages with YUM or DNF

The command to install a package is dnf install. Installing a package requires administrative privileges, so you need to remember to run all such commands with sudo.

When you run an installation command, YUM and DNF will automatically download package index files from the repository, so you don't need to take any action to ensure that you have up-to-date repository metadata. A disadvantage of that approach is that sometimes you may have to wait for it to download that metadata, which may take a few minutes or more on very slow connections.

However, the installation process is completely automated. YUM and DNF will show you a list of packages that would be downloaded and installed and how much space they would need. If you say yes, it will proceed with the download and installation:

```
$ sudo dnf install hello
Last metadata expiration check: 0:44:59 ago on Mon 17 Oct 2022
10:06:44 AM IST.
Dependencies resolved.
================================================================
 Package        Architecture  Version              Repository     Size
================================================================
Installing:
 hello          x86_64        2.10-7.fc36          fedora         70 k

Transaction Summary
================================================================
Install  1 Package

Total download size: 70 k
Installed size: 179 k
Is this ok [y/N]: y
Downloading Packages:
hello-2.10-7.fc36.x86_64.rpm          278 kB/s |  70 kB      00:00
----------------------------------------------------------------
Total                                 147 kB/s |  70 kB      00:00
Running transaction check
Transaction check succeeded.
Running transaction test
Transaction test succeeded.
Running transaction
Preparing        :                                             1/1
Installing       : hello-2.10-7.fc36.x86_64                    1/1
Running scriptlet: hello-2.10-7.fc36.x86_64                    1/1
Verifying        : hello-2.10-7.fc36.x86_64              1/1
Installed:
  hello-2.10-7.fc36.x86_64
Complete!
```

Removing a package is also straightforward. Just run `dnf remove hello`.

Unlike RPM, DNF and YUM do have a concept of protected packages, so they will outright refuse to delete essential packages such as `bash` or `rpm`:

```
$ sudo dnf remove bash
Error:
 Problem: The operation would result in removing the following
protected packages: dnf
(try to add '--skip-broken' to skip uninstallable packages)
```

This protection from erroneous user actions is a significant reason to use YUM or DNF to remove packages despite the fact that RPM is technically sufficient for that task.

Installing and removing packages with APT

APT uses a different approach to metadata download: it leaves that operation to the user. Unlike YUM and DNF, APT will never automatically download repository indices, so on a freshly deployed Debian-based system, trying to search for or install packages will give you a *package not found* error.

> **Note**
>
> APT consists of sub-tools for different purposes: `apt-get` for repository index updates and package actions (such as installing, removing, or listing package files), and `apt-cache` for package search. In modern Debian versions, you will find a combined tool named `apt`. You may also want to use an alternative frontend named `aptitude`, which we will discuss later.

Before you can install anything, you need to force the repository index download with `apt-get update` or `apt update`. `update` in its name refers only to a metadata update, not to a package update:

```
$ sudo apt-get update
Get:1 http://deb.debian.org/debian buster InRelease [122 kB]
Get:2 http://security.debian.org/debian-security buster/updates
InRelease [34.8 kB]

                                          Get:3 http://deb.debian.
org/debian buster-updates InRelease [56.6 kB]
...
```

After that, you can search for packages with `apt-cache search` or `apt search` and install them with `apt install` or `apt-get install`.

For example, with the following command, you can install the `hello` package automatically from the repositories, without having to download it by hand:

```
$ sudo apt install hello
Reading package lists... Done
Building dependency tree
Reading state information... Done
The following NEW packages will be installed:
  hello
0 upgraded, 1 newly installed, 0 to remove and 124 not upgraded.
Need to get 56.1 kB of archives.
After this operation, 287 kB of additional disk space will be used.
Get:1 http://deb.debian.org/debian buster/main amd64 hello amd64
2.10-2 [56.1 kB]
Fetched 56.1 kB in 0s (1,385 kB/s)
Selecting previously unselected package hello.
(Reading database ... 113404 files and directories currently
installed.)
Preparing to unpack .../hello_2.10-2_amd64.deb …
Unpacking hello (2.10-2) …
Setting up hello (2.10-2) …
Processing triggers for man-db (2.8.5-2) …
$ hello
Hello, world!
```

APT also includes protection against the removal of essential packages. Trying to remove `bash`, for example, will require a special confirmation:

```
$ sudo apt remove bash
Reading package lists... Done
Building dependency tree
Reading state information... Done
The following packages will be REMOVED:
  bash
WARNING: The following essential packages will be removed.
This should NOT be done unless you know exactly what you are doing!
  bash
0 upgraded, 0 newly installed, 1 to remove and 124 not upgraded.
After this operation, 6,594 kB disk space will be freed.
You are about to do something potentially harmful.
To continue type in the phrase 'Yes, do as I say!'
 ?]
```

Needless to say, you should not do this unless you are absolutely certain that what you are doing is safe—for example, if you have made sure that every user uses a shell other than Bash and there are no scripts in the system that require Bash rather than a POSIX Bourne shell.

Removing a non-essential package, such as `hello`, will not raise any such errors:

```
$ sudo apt-get remove hello
Reading package lists... Done
Building dependency tree... Done
Reading state information... Done
The following packages will be REMOVED:
  hello
0 upgraded, 0 newly installed, 1 to remove and 0 not upgraded.
After this operation, 284 kB disk space will be freed.
Do you want to continue? [Y/n] Y
(Reading database ... 72242 files and directories currently
installed.)
Removing hello (2.10-3) ...
$ hello
bash: /usr/bin/hello: No such file or directory
```

There is also the `apt-get purge` command, which removes not only executables and data files but also all configuration files associated with the package. Most of the time, `apt-get remove` is sufficient, but to remove a package such as a web server, you may prefer `apt-get purge` instead.

Searching for packages

Both APT and YUM/DNF provide a search command, so if you don't know the exact name of a package, you can search for it. However, they search for a pattern in every field but only display the name and the short description by default, so search results may look odd and include entries that seem to have nothing to do with your request.

For example, let's try to search for the `hello` package on a Debian system:

```
$ apt-cache search hello
junior-system - Debian Jr. System tools
elpa-find-file-in-project - quick access to project files in Emacs
python-flask - micro web framework based on Werkzeug and Jinja2 -
Python 2.7
...
libghc-lambdabot-misc-plugins-prof - Lambdabot miscellaneous plugins;
profiling libraries
hello - example package based on GNU hello
hello-traditional - example package not using any helper package
iagno - popular Othello game for GNOME
```

Sometimes, you may want to search for a package that provides a certain command or a library file (for example, if you are getting a script error that complains that it wasn't found).

In YUM and DNF, there is a built-in option for that: `whatprovides`. It supports both exact file paths and wildcard matches. Suppose you want to install a package that provides a `hello` command. Executable files of commands are always in some `bin/` directory, but we don't know whether it's going to be `/bin`, `/usr/bin`, `/usr/sbin`, or something else. However, we can search for `*/bin/hello` to find any executable with that name in any package. It will include some irrelevant results, but it will tell us what we want to know:

```
$ dnf whatprovides '*/bin/hello'
hello-2.10-7.fc36.x86_64 : Prints a familiar, friendly greeting
Repo         : fedora
Matched from:
Filename     : /usr/bin/hello
rr-testsuite-5.5.0-3.fc36.x86_64 : Testsuite for checking rr
functionality
Repo         : fedora
Matched from:
Filename     : /usr/lib64/rr/testsuite/obj/bin/hello
rr-testsuite-5.6.0-1.fc36.x86_64 : Testsuite for checking rr
functionality
Repo         : updates
Matched from:
Filename     : /usr/lib64/rr/testsuite/obj/bin/hello
```

In Debian-based systems, it's not that simple. You will need to install an optional `apt-file` tool (`apt-get install apt-file`) and run `apt-file update` to download additional indices.

It also doesn't support wildcard matching, so if you don't know the exact path, you will need to supply a Perl-compatible regular expression for your search:

```
$ apt-file search --regexp '(.*)/bin/hello'
Searching, found 10 results so far ...
hello: /usr/bin/hello
hello-traditional: /usr/bin/hello
libpion-dev: /usr/bin/helloserver
pvm-examples: /usr/bin/hello.pvm
pvm-examples: /usr/bin/hello_other
```

As you can see, YUM and DNF provide more functionality out of the box, while APT may require installing additional packages. Nonetheless, it should be possible to execute the same search operations with all of those tools.

System upgrade

Another advantage of high-level package managers is that they automate the upgrade of the entire system (or at least packages installed from repositories).

Upgrading a system with YUM or DNF

The command to upgrade all packages is dnf upgrade or yum upgrade. To force a repository index download, you can add --refresh. In some cases, you will also want to remove conflicting or outdated packages to complete an upgrade; in that case, you may need dnf upgrade --allowerasing:

```
$ sudo dnf upgrade
Last metadata expiration check: 1:33:10 ago on Mon 17 Oct 2022
10:06:44 AM IST.
Dependencies resolved.
================================================================
 Package             Arch    Version          Repo      Size
================================================================
Installing:
 kernel              x86_64 5.19.15-201.fc36    updates  264 k
 kernel-core         x86_64 5.19.15-201.fc36    updates   49 M
 kernel-devel        x86_64 5.19.15-201.fc36    updates   16 M
 kernel-modules      x86_64 5.19.15-201.fc36    updates   58 M
 kernel-modules-extra x86_64 5.19.15-201.fc36   updates  3.7 M
Upgrading:
 amd-gpu-firmware    noarch 20221012-141.fc36   updates   15 M
 ansible-srpm-macros noarch 1-8.1.fc36          updates  8.5 k
 appstream           x86_64 0.15.5-1.fc36       updates  668 k
 bash                x86_64 5.2.2-2.fc36         updates  1.8 M
 ...
Installing dependencies:
  iirl              x86_64 1.9.3-1.fc36         updates   27 k
Removing:
 kernel              x86_64 5.19.11-200.fc36    @updates    0
 kernel-core         x86_64 5.19.11-200.fc36    @updates   92 M
 kernel-devel        x86_64 5.19.11-200.fc36    @updates   65 M
 kernel-modules      x86_64 5.19.11-200.fc36    @updates   57 M
 kernel-modules-extra x86_64 5.19.11-200.fc36   @updates  3.4 M
Removing dependent packages:
  kmod-VirtualBox-5.19.11-200.fc36.x86_64
                    x86_64 6.1.38-1.fc36      @@commandline
                                                          160 k
```

```
Transaction Summary
================================================================
Install    6 Packages
Upgrade   77 Packages
Remove     6 Packages
Total download size: 433 M
Is this ok [y/N]:
```

If you say `yes`, it will automatically download new package versions and overwrite old packages with them.

Upgrading a Fedora system to a new distribution version requires a different procedure, however. This is only possible with DNF and needs a plugin available from Fedora repositories. The command sequence to upgrade an older system to Fedora 36 (the current version in 2022) would be as follows:

```
$ sudo dnf upgrade --refresh
$ sudo dnf install dnf-plugin-system-upgrade
$ sudo dnf system-upgrade download --releasever=36
$ sudo dnf system-upgrade reboot
```

The `dnf system-upgrade download --releasever=36` command will download all packages required to upgrade to a new version and run a transaction check. On rare occasions, you will need to remove certain packages if they are no longer available in a new Fedora version. If the check is successful, you can start the upgrade procedure with `dnf system-upgrade reboot`—your system will boot into a minimal environment to perform the upgrade, then boot to the new Fedora version as normal.

Upgrading a system with APT

APT includes upgrade commands with different functionality instead of using modifier options as YUM and DNF do:

- `apt upgrade` and `apt-get upgrade` will only upgrade installed packages to newer versions if they are available, but never perform any other actions
- `apt full-upgrade` or `apt-get dist-upgrade` may remove packages if it's required to upgrade the system as a whole

Most of the time, you should use `apt-get dist-upgrade` because when you update packages within the same distribution version, package removal events are incredibly rare, while if you upgrade to a new distribution version, there will not be a way around it—you will need to have them removed one way or another before you can upgrade.

This is what a typical package update will look like:

```
$ sudo apt-get dist-upgrade
Reading package lists... Done
Building dependency tree
Reading state information... Done
Calculating upgrade... Done
The following package was automatically installed and is no longer
required:
  linux-image-4.19.0-10-amd64
Use 'sudo apt autoremove' to remove it.
The following NEW packages will be installed:
  linux-headers-4.19.0-22-amd64 linux-headers-4.19.0-22-common linux-
image-4.19.0-22-amd64
The following packages will be upgraded:
  base-files bind9-host bzip2 curl ... open-vm-tools openjdk-11-jre
openjdk-11-jre-headless
  openssl publicsuffix python-paramiko qemu-utils rsyslog tzdata unzip
vim vim-common vim-runtime vim-tiny xxd xz-utils zlib1g zlib1g-dev
124 upgraded, 3 newly installed, 0 to remove and 0 not upgraded.
Need to get 186 MB of archives.
After this operation, 330 MB of additional disk space will be used.
Do you want to continue? [Y/n]
```

Upgrading a Debian-based system to a new distribution version is more involved. You will need to look up the codename of the new version and replace the old one in your repository configuration files. For example, if you are upgrading from Debian 10 to 11, you need to replace every occurrence of buster (the codename of Debian 10) with bullseye (the codename of Debian 11), then run apt-get dist-upgrade and reboot your system.

As you can see, system-wide upgrade procedures are conceptually similar in Red Hat and Debian-based distributions, even if the exact commands and implementation details differ.

Summary

In this chapter, we learned about low-level and high-level package managers used by popular Linux distributions. We learned how to install and remove packages using DNF and APT and how to perform a system upgrade. We also learned how to inspect package files by hand and interpret their internal structure—while that task is much less frequent, it's important to know to have a deeper understanding of software packaging and the package management process. However, package managers offer a lot of additional options and capabilities, so make sure to read their documentation.

In the next chapter, we will learn about network configuration and debugging in Linux systems.

Further reading

- RPM documentation: `http://rpm.org/documentation.html`
- dpkg: `https://www.dpkg.org/`
- DNF documentation: `https://dnf.readthedocs.io/en/latest/`
- APT documentation: `https://wiki.debian.org/PackageManagement`

9

Network Configuration and Debugging

All modern systems are networked, so network configuration and troubleshooting are fundamental skills for every systems administrator. In this chapter, we will learn how the Linux network stack works and how to use the tools for working on it—both universal and distribution-specific.

In this chapter, we will cover the following topics:

- Linux network stack
- Network interfaces and addresses in Linux
- Routes and neighbor tables
- NetworkManager
- Distribution-specific configuration methods
- Network troubleshooting

Linux network stack

To end users who only interact with the network through applications and only configure network access through a graphical user interface, the network stack of their operating system looks like a single abstraction. However, for administrators, it is important to understand its structure because different parts of the stack are implemented by different software and administered by different tools.

This contrasts Linux distributions with many proprietary operating systems where most network functions are built-in and cannot be replaced individually. In a Linux distribution, performance-critical functionality is implemented by the Linux kernel itself, but many other functions, such as the dynamic configuration of IP addresses and routes through **Dynamic Host Configuration Protocol (DHCP)**, are done by third-party tools, and there can be multiple competing implementations.

There are also different tools for managing the network functionality of the Linux kernel. The kernel allows userspace processes to retrieve and change its network configuration via *Netlink* protocol sockets and, technically, anyone can write a tool for managing IP addresses and routes. In practice, there are two suites of network administration tools: the legacy tools (`ifconfig`, `vconfig`, `route`, `brctl`, and so on), which are only kept for compatibility and do not support many new features of the kernel network stack, and the modern `iproute2` toolkit, which provides access to all kernel functionality via `ip` and `tc` utilities.

Sometimes, more than one implementation of the same or similar functionality may exist in the kernel as well. One prominent example is the Netfilter firewall subsystem, which currently includes the older `iptables` framework and the newer `nftables` implementation.

Different implementations of userspace tools may also be either legacy implementations that are being gradually replaced with newer alternatives, or there can also be multiple alternative implementations with different design goals. For example, in 2022, the ISC DHCP server was declared unsupported by its maintainers who went on to work on the newer Kea project. However, Kea is not the only alternative to the ISC DHCP server. Some people may want to switch to other projects instead. For example, small networks can benefit from `dnsmasq`, which combines a DHCP server with DNS forwarding and other functions, which is especially useful for running it on small office routers with limited hardware resources.

Some of the most commonly used Linux network functionality backends and management tools are summarized in the following table:

Component	Implementation(s)	Tools
Ethernet	Linux kernel	Network card settings tweaking: `ethtool` MAC address settings, VLANs: `iproute2` (modern); `vconfig` (legacy)
Wi-Fi (framing and forwarding)	Linux kernel	`iw`
Wi-Fi (authentication and access point functionality)	`hostapd`	
IPv4 and IPv6 routing	Linux kernel	`iproute2` (modern) `ifconfig`, `route`, etc. (legacy)

Component	Implementation(s)	Tools
Bridging (software switch)	Linux kernel	`iproute2` (modern); `brctl` (legacy)
QoS and traffic shaping	Linux kernel	`tc` (part of `iproute2`)
IPsec (packet encryption and checksum calculation)	Linux kernel	`iproute2`
IPsec (IKE session management)	strongSwan, Openswan, Raccoon (legacy)...	
DHCP client	ISC DHCP, `dhcpcd`	
DHCP server	ISC DHCP, ISC Kea, `dnsmasq`	

Table 9.1 – Linux network stack components

Finally, there are high-level management tools such as NetworkManager that tie multiple tools and components under a single user interface. Let's learn about the kernel parts of the network stack and how to manage them with `iproute2` first. Then, we will see how to simplify and automate that with high-level tools in different distributions.

Network interfaces and addresses in Linux

Network interface is a generic term for physical and virtual network links that can carry addresses. The correspondence between physical network cards and network interfaces as the kernel sees them is not one-to-one. A network card with four ports is a single PCI device, but every one of its ports is seen as a separate link by the kernel.

There are also virtual links. Some virtual links are tied to physical network ports, but other types are fully independent. For example, MACVLAN links allow administrators to send traffic from certain IP addresses using a different MAC address. Since an Ethernet interface by definition cannot have multiple MAC addresses, Linux solves that problem by creating virtual interfaces on top of a physical Ethernet port and assigning different MAC and IP addresses to it. Multiplexing Ethernet traffic using 802.1Q VLAN or 802.3ad QinQ (nested VLAN) is also done by creating a virtual link that is bound to a specific underlying link.

However, interfaces for tunneling protocols such as IPIP and GRE are not tied to any underlying links. They require tunnel endpoint addresses, but those addresses can belong to any interface. There are also dummy interfaces that are used either for local process communication or for assigning addresses that must be reachable through any physical interface:

Link type	Relationship with physical devices
Ethernet, Wi-Fi	Associated with physical cards or ports on those cards
802.1Q VLAN, 802.3ad QinQ, MACVLAN	Tied to a single physical link
IPIP, GRE, dummy	Purely virtual

Table 9.2 – Network link types and their relationships with physical devices

In the following sections, we will learn how to retrieve information about network interfaces and configure them.

Discovering physical network devices

Discovering all physical network devices in a Linux system can be a challenging task. They can be connected to different buses, including PCI and USB, and those buses use different device class identifiers.

Consider the following PCI device listing from a laptop:

```
$ lspci
00:00.0 Host bridge: Intel Corporation 11th Gen Core Processor Host
Bridge/DRAM Registers (rev 01)
00:02.0 VGA compatible controller: Intel Corporation TigerLake-LP GT2
[Iris Xe Graphics] (rev 01)
00:04.0 Signal processing controller: Intel Corporation TigerLake-LP
Dynamic Tuning Processor Participant (rev 01)
00:06.0 PCI bridge: Intel Corporation 11th Gen Core Processor PCIe
Controller (rev 01)
...
00:14.3 Network controller: Intel Corporation Wi-Fi 6 AX201 (rev 20)
...
02:00.0 Non-Volatile memory controller: Samsung Electronics Co Ltd
NVMe SSD Controller 980
03:00.0 Ethernet controller: Realtek Semiconductor Co., Ltd.
RTL8111/8168/8411 PCI Express Gigabit Ethernet Controller (rev 15)
```

Network devices are easy to identify visually here. There is a Wi-Fi controller (00:14.3) and an Ethernet card (03:00.0). Automatically filtering out everything but network devices from that listing is a bit trickier. We can use the fact that the PCI class for network devices is 02xx, and there is a way to include device class numbers in the output with lspci -nn:

```
$ lspci -nn | grep -E '\[02[0-9]+\]'
00:14.3 Network controller [0280]: Intel Corporation Wi-Fi 6 AX201
[8086:a0f0] (rev 20)
03:00.0 Ethernet controller [0200]: Realtek Semiconductor Co., Ltd.
```

```
RTL8111/8168/8411 PCI Express Gigabit Ethernet Controller [10ec:8168]
(rev 15)
```

Given that, to see all network devices you will need to look in both PCI and USB device listings, it is better to use high-level third-party tools such as `lshw`. With the `lshw -class` command you can view all available network devices in one step: both wired and wireless, connected to any buses. It also shows a lot of additional information about devices:

```
$ sudo lshw -class network
  *-network
        description: Wireless interface
        product: Wi-Fi 6 AX201
        vendor: Intel Corporation
        physical id: 14.3
        bus info: pci@0000:00:14.3
        logical name: wlp0s20f3
        version: 20
        serial: 12:15:81:65:d2:2e
        width: 64 bits
        clock: 33MHz
        capabilities: pm msi pciexpress msix bus_master cap_list
ethernet physical wireless
        configuration: broadcast=yes driver=iwlwifi
driverversion=5.19.16-200.fc36.x86_64 firmware=71.058653f6.0 QuZ-a0-
jf-b0-71.u latency=0 link=no multicast=yes wireless=IEEE 802.11
        resources: iomemory:600-5ff irq:16 memory:6013038000-601303bfff
  *-network
        description: Ethernet interface
        product: RTL8111/8168/8411 PCI Express Gigabit Ethernet
Controller
        vendor: Realtek Semiconductor Co., Ltd.
        physical id: 0
        bus info: pci@0000:03:00.0
        logical name: enp3s0
        version: 15
        serial: 60:18:95:39:ca:d0
        capacity: 1Gbit/s
        width: 64 bits
        clock: 33MHz
        capabilities: pm msi pciexpress msix bus_master cap_list
ethernet physical tp mii 10bt 10bt-fd 100bt 100bt-fd 1000bt-fd
autonegotiation
        configuration: autonegotiation=on broadcast=yes driver=r8169
driverversion=5.19.16-200.fc36.x86_64 firmware=rtl8168h-2_0.0.2
02/26/15 latency=0 link=no multicast=yes port=twisted pair
```

```
      resources: irq:16 ioport:3000(size=256) memory:72004000-
72004fff memory:72000000-72003fff
```

As you can see, the lshw output also includes logical interface names rather than just bus addresses. Every network interface in Linux has a unique name, but their names are not completely determined by their hardware type and bus port. Let us examine the issue of interface naming in more detail.

Network interface names

The Linux kernel does not ascribe any special significance to network interface names. In some operating systems, interface names are completely determined by the kernel so that the first Ethernet device might always be named Ethernet0, and there is no way for an administrator to change that. In Linux, that is not the case, and names can be arbitrary. In fact, most distributions include a userspace helper for renaming network interfaces at boot time according to a default policy or custom configuration. Formerly, the most common helper was udev. Now, it is usually systemd-udevd.

Historically, Ethernet devices were named ethX by default, as per the kernel's built-in naming scheme. By the 2020s, most distributions switched to systemd for service management and adopted its **predictable network interface names** scheme as their default option.

The issue with the original naming scheme is that the kernel's device probing is not deterministic, so in some situations, especially when new network cards were added or old cards were removed, old names could be assigned to different physical devices (for example, a card formerly named eth2 would become eth1). On the other hand, if a machine had a single network interface, it was guaranteed to be named eth0.

The naming scheme of systemd is predictable in the sense that network interface names are guaranteed to stay the same across reboot and hardware changes. The price for that is that users and scriptwriters cannot make any assumptions about names. Even if a machine only has a single network card, it can be named, for example, eno1 (Ethernet network, onboard, number 1) or enp3s0 (Ethernet network, PCI, slot 3:0).

It is possible to switch to the original naming scheme, either by adding net.ifnames=0 to the kernel command line in the GRUB configuration, or by executing the following command:

```
$ sudo ln -s /dev/null /etc/systemd/network/99-default.link
```

It is also possible to permanently assign custom names to certain network interfaces by creating systemd link files.

Using the ip command

In modern Linux distributions, all network discovery and setup are done either with utilities from the iproute2 package or with high-level tools such as NetworkManager. We will omit the legacy tools such as ifconfig and focus on the ip utility from iproute2.

Even though that utility is named `ip`, its functionality is much broader, and it provides an interface to all features of the kernel network stack that are related to network interfaces, addresses, and routing.

One thing to note is that in some distributions such as Fedora, that utility may be installed in /sbin or /usr/sbin—directories meant for administrative tools and often absent from the $PATH environment variable in shell configurations for unprivileged users. Thus, attempts to execute it from an unprivileged shell will result in a command not found error even though `iproute2` is installed. In that case, you may want to either add /sbin to your $PATH or always run `sudo ip` instead. Commands that change network settings indeed require administrative privileges but commands for viewing them usually do not.

Note that changes you make with `ip` only remain active until the next reboot and permanent changes must be made in distribution-specific configuration files instead or added to a script executed at boot time. If you are experimenting on a desktop or a laptop computer with NetworkManager running, then it may also override your changes on, for example, Wi-Fi reconnects.

Discovering and inspecting logical links

To view all network interfaces, both physical and virtual, you can run `ip link list` or simply `ip link`. Note that `ip` allows abbreviating subcommands and options, so you can also write `ip li li`, but we will use full forms throughout the chapter for better readability:

```
$ ip link list
1: lo: <LOOPBACK,UP,LOWER_UP> mtu 65536 qdisc noqueue state UNKNOWN
mode DEFAULT group default qlen 1000
    link/loopback 00:00:00:00:00:00 brd 00:00:00:00:00:00
2: ens192: <BROADCAST,MULTICAST,UP,LOWER_UP> mtu 1500 qdisc mq state
UP mode DEFAULT group default qlen 1000
    link/ether 00:50:56:91:a2:b6 brd ff:ff:ff:ff:ff:ff
```

In this output from a VM, we see the loopback device (`lo`) and a single Ethernet card named according to the systemd predictable network interface naming convention (`ens192`).

The loopback device is present in every Linux system. Its role is to enable communication between local processes over IP and it carries addresses designated for that use: `127.0.0.1/8` for IPv4 and `::1/128` for IPv6.

The output for the `ens192` Ethernet device has more data. In the `link/ether` field, you can see its MAC address (`00:50:56:91:a2:b6`).

You may also notice seemingly redundant `<...UP,LOWER_UP>` and `state UP` bits in the output. They refer to different facts about that network interface: `UP` inside the angle brackets tells us that the link is not intentionally disabled by the administrator, while `state UP` refers to the actual state (whether it is connected to any other network device or not—with a physical cable or a virtual link, in the case of VMs).

To illustrate the distinction, let us examine a physical network on another machine that is not connected to anything. To view information about a single link, you can use ip link show <name>:

```
$ /sbin/ip link show enp3s0
2: enp3s0: <NO-CARRIER,BROADCAST,MULTICAST,UP> mtu 1500 qdisc fq_codel
state DOWN mode DEFAULT group default qlen 1000
    link/ether 60:18:95:39:ca:d0 brd ff:ff:ff:ff:ff:ff
```

As you can see, in the angle brackets its state is UP, but state DOWN tells us that it is not active, and NO-CARRIER explains why—it is disconnected (an Ethernet link can also be down despite being connected to something, for example, due to a settings mismatch with the other side).

Now let's disable a link to see what an intentionally disabled link looks like. You can do it with sudo ip link set dev <name> down:

```
$ ip link show eth2
6: eth2: <BROADCAST,MULTICAST,UP,LOWER_UP> mtu 1500 qdisc pfifo_fast
state UP mode DEFAULT group default qlen 1000
    link/ether 00:50:56:9b:18:ed brd ff:ff:ff:ff:ff:ff
$ sudo ip link set dev eth2 down
$ ip link show eth2
6: eth2: <BROADCAST,MULTICAST> mtu 1500 qdisc pfifo_fast state DOWN
mode DEFAULT group default qlen 1000
    link/ether 00:50:56:9b:18:ed brd ff:ff:ff:ff:ff:ff
```

You can see that when the link was taken down, the UP token disappeared from the part inside the angle brackets, and also its state field changed to DOWN.

Viewing and changing Ethernet link MAC addresses

Every Ethernet and Wi-Fi card has a globally unique, burnt-in MAC address. To make sure that no two network devices will ever conflict if they are connected to the same network, manufacturers request blocks of MAC addresses and keep track of every MAC address they assign to their hardware products so that no address is ever assigned twice.

However, end users may have reasons to assign their own MAC address to a network interface. For example, many internet service providers register the first MAC address of the subscriber's router port and then require all future connection attempts to use the same address. If the user replaces or upgrades the router (or a network card in it), it is often easier to just assign the original port's MAC address than ask the ISP support to update their records. You can change the MAC address (until the next reboot) with ip link set dev <name> address <MAC>:

```
$ ip link show enp3s0
2: enp3s0: <NO-CARRIER,BROADCAST,MULTICAST,UP> mtu 1500 qdisc fq_codel
state DOWN mode DEFAULT group default qlen 1000
    link/ether 60:18:95:39:ca:d0 brd ff:ff:ff:ff:ff:ff
```

```
$ sudo ip link set dev enp3s0 address de:ad:be:ef:ca:fe
$ /sbin/ip link show enp3s0
2: enp3s0: <NO-CARRIER,BROADCAST,MULTICAST,UP> mtu 1500 qdisc fq_codel
state DOWN mode DEFAULT group default qlen 1000
    link/ether de:ad:be:ef:ca:fe brd ff:ff:ff:ff:ff:ff permaddr
60:18:95:39:ca:d0
```

Note that ip link show displays the new, manually assigned MAC address now (de:ad:be:ef:ca:fe).

While ip only shows the MAC address that the kernel uses for sending Ethernet frames, it's possible to retrieve the default, burnt-in address with ethtool instead. You can use either ethtool --show-permaddr or its short version, ethtool -P, as follows:

```
$ ethtool --show-permaddr enp3s0
Permanent address: 60:18:95:39:ca:d0
$ ethtool -P enp3s0
Permanent address: 60:18:95:39:ca:d0
```

It is useful to know how to change MAC addresses even though it is not a very common task. Next, we will learn how to manage IP addresses.

Viewing and changing IP addresses

Commands for viewing and changing IP addresses are similar to those for links and MAC addresses but use the address command family instead of link. To view addresses on all links, you can run ip address show, just ip address, or an abbreviated version—ip a:

```
$ ip address show
1: lo: <LOOPBACK,UP,LOWER_UP> mtu 65536 qdisc noqueue state UNKNOWN
group default qlen 1000
    link/loopback 00:00:00:00:00:00 brd 00:00:00:00:00:00
    inet 127.0.0.1/8 scope host lo
        valid_lft forever preferred_lft forever
    inet6 ::1/128 scope host
        valid_lft forever preferred_lft forever
2: ens192: <BROADCAST,MULTICAST,UP,LOWER_UP> mtu 1500 qdisc mq state
UP group default qlen 1000
    link/ether 00:50:56:91:a2:b6 brd ff:ff:ff:ff:ff:ff
    inet 10.217.40.163/24 brd 10.217.40.255 scope global dynamic
ens192
        valid_lft 58189sec preferred_lft 58189sec
    inet6 fe80::250:56ff:fe91:a2b6/64 scope link
        valid_lft forever preferred_lft forever
```

You can also limit the output to just one interface, such as in ip address show lo.

The output of `ip address show` includes MAC addresses for Ethernet and other data link layer interfaces, so often, you can use that command instead of `ip link list`, unless you specifically want to exclude IP addresses from the output.

It is possible to show only IPv4 or only IPv6 addresses with the `-4` and `-6` options. We can demonstrate it on the loopback interface (`lo`) since it is guaranteed to have both IPv4 and IPv6 addresses (unless IPv6 is disabled explicitly):

```
$ ip -4 address show lo
1: lo: <LOOPBACK,UP,LOWER_UP> mtu 65536 qdisc noqueue state UNKNOWN
group default qlen 1000
    inet 127.0.0.1/8 scope host lo
       valid_lft forever preferred_lft forever
$ ip -6 address show lo
1: lo: <LOOPBACK,UP,LOWER_UP> mtu 65536 state UNKNOWN qlen 1000
    inet6 ::1/128 scope host
       valid_lft forever preferred_lft forever
```

Now let's see how to add and remove addresses. For safe experiments that knowingly will not affect any network interface used for real traffic, we will create a dummy interface. Dummy interfaces are conceptually similar to the loopback interface. However, there can be multiple dummy interfaces in the same system, while there can only be one loopback (this situation is unique to Linux; many other operating systems allow multiple loopback interfaces instead of using a different interface type):

```
$ sudo ip link add name dummy1 type dummy
$ sudo ip link set dev dummy1 up
$ ip link list type dummy
16: dummy1: <BROADCAST,NOARP,UP,LOWER_UP> mtu 1500 qdisc noqueue state
UNKNOWN mode DEFAULT group default qlen 1000
    link/ether 9a:9c:10:42:a6:ea brd ff:ff:ff:ff:ff:ff
```

All virtual interfaces are created in the DOWN state in Linux, so we brought the dummy1 link up by hand. Now it is ready for experiments with addresses.

You can assign an address with `ip address add <addr> dev <name>`. Just as with MAC addresses, such changes will not survive reboots, so this method is only good for experiments and troubleshooting sessions. We will use addresses from `192.0.2.0/24`—a network reserved for examples and documentation:

```
$ sudo ip address add 192.0.2.1/24 dev dummy1
$ ip address show dev dummy1
16: dummy1: <BROADCAST,NOARP,UP,LOWER_UP> mtu 1500 qdisc noqueue state
UNKNOWN group default qlen 1000
    link/ether 9a:9c:10:42:a6:ea brd ff:ff:ff:ff:ff:ff
    inet 192.0.2.1/24 scope global dummy1
       valid_lft forever preferred_lft forever
```

```
        inet6 fe80::989c:10ff:fe42:a6ea/64 scope link
            valid_lft forever preferred_lft forever
```

Note that executing `ip address add` for the second time with a different address will not replace the old address but rather add a second address. There is no limit on the number of addresses on a single network interface in Linux; you can assign as many as you want. If you want to replace an address, you can add the new one first and then remove the old one. Let's see how we can replace `192.0.2.1` with `192.0.2.2`:

```
$ sudo ip address add 192.0.2.2/24 dev dummy1
$ sudo ip address show dummy1
16: dummy1: <BROADCAST,NOARP,UP,LOWER_UP> mtu 1500 qdisc noqueue state
UNKNOWN group default qlen 1000
    link/ether 9a:9c:10:42:a6:ea brd ff:ff:ff:ff:ff:ff
    inet 192.0.2.1/24 scope global dummy1
        valid_lft forever preferred_lft forever
    inet 192.0.2.2/24 scope global secondary dummy1
        valid_lft forever preferred_lft forever
$ sudo ip address delete 192.0.2.1/24 dev dummy1
$ sudo ip address show dummy1
16: dummy1: <BROADCAST,NOARP,UP,LOWER_UP> mtu 1500 qdisc noqueue state
UNKNOWN group default qlen 1000
    link/ether 9a:9c:10:42:a6:ea brd ff:ff:ff:ff:ff:ff
    inet 192.0.2.2/24 scope global dummy1
        valid_lft forever preferred_lft forever
```

It is also possible to remove all addresses from a network interface at once using `sudo ip address flush dev <name>`.

Most of the time, you will configure IP addresses using high-level configuration tools that we will discuss later in this chapter. However, knowing those commands can help you verify address configuration and change network interface addresses temporarily during troubleshooting sessions or emergency configuration changes.

Routes and neighbor tables

To be able to communicate with other hosts over the network, it's not enough for a host to have an address. It also needs to know how to reach other hosts. Modern networks use layered protocol stacks, and the Linux kernel is responsible for the **Data Link** and **Network** layers according to the OSI model.

At the data link layer, there are protocols such as Ethernet and Wi-Fi — both are multiple-access broadcast networks and require dynamic discovery of neighbors in the same network segment. At the data link layer, hosts are identified by their MAC addresses. Direct communication at the data link layer is only possible within the same segment. If a network layer protocol packet encapsulated in a data link layer protocol frame must travel further, it's extracted from the original frame and encapsulated in a new one.

Above the data link layer are network layer protocols—IPv4 and IPv6. IP packets can be sent either to hosts in the same data link layer segment or routed to other networks and may traverse many data link layer connections on their way to their destination.

However, to be able to send an IP packet to another host or a router, the kernel needs to build an association between the IP addresses of those machines and their MAC addresses and maintain tables of such associations.

ARP and NDP neighbor tables

IPv4 and IPv6 protocols share many similarities but use different neighbor discovery mechanisms. The older IPv4 protocol uses **Address Resolution Protocol** (**ARP**) to determine the MAC addresses of hosts with given IP addresses. ARP was not designed to be extensible, and the switch from 32-bit addresses in IPv4 to 128-bit ones in IPv6 required the development of a new protocol, so its designers used it as a chance to revise many old assumptions and add many new features. The result was named **Neighbor Discovery Protocol** (**NDP**), and, unlike ARP, it allows hosts to discover routers and dynamically configure public addresses, and detect address conflicts.

To view the ARP table, you can run `ip -4 neighbor show`. You can also shorten it to just `ip -4 neighbor` or `ip -4 neigh`. Note that those commands also support the British spelling (`neighbour`) if you prefer to use it. If you omit `-4` or `-6`, that command will show entries for both protocols, so if your system does not have IPv6 configured or if you do not want to filter, you can omit `-4`:

```
$ ip -4 neighbor show
10.217.32.199 dev eth1 lladdr 00:0c:29:62:27:03 REACHABLE
10.217.32.132 dev eth1   FAILED
10.217.32.111 dev eth1   FAILED
10.217.32.102 dev eth1 lladdr 00:50:56:85:d9:72 DELAY
10.217.32.201 dev eth1 lladdr 00:50:56:9b:dd:47 STALE
10.217.32.99 dev eth1 lladdr 00:0c:29:5f:92:1d REACHABLE
10.217.32.202 dev eth1 lladdr 00:50:56:9b:bc:28 STALE
10.217.32.117 dev eth1 lladdr 00:50:56:9b:7e:e3 REACHABLE
```

It is also possible to filter the output and only show entries for one network interface, for example, with `ip -4 neighbor show dev eth1`.

The field for MAC addresses is called the **Link-Layer Address** (**lladdr**). The reason is that the neighbor table output format is the same for multiple data link layer protocols that may not name their link-layer addresses MAC addresses. There are also situations when the link-layer address for an IPv4 host is itself an IPv4 address—that's how multipoint GRE tunnels work, for example (it's part of the dynamic multipoint VPN technology that also includes IPsec for encryption and the next-hop resolution protocol for neighbor discovery).

Every association is a triple rather than a pair: MAC address, IPv4 address, and network interface. Every Ethernet interface belongs to its own data link layer segment, so to send an IP packet correctly, the system needs to know which network card to send it from. MAC addresses must only be unique within the same segment.

It is possible to only show entries with a specific state. For example, this is how to view only address associations that have been recently discovered or confirmed:

```
ip neighbor show dev <name> nud reachable
```

Stale entries eventually disappear from the table. When an IP address is moved to a machine with a different MAC address, the kernel will also eventually discover it. But if waiting is not an option and an IP address must be moved with minimum downtime, you can manually remove an entry and force a new ARP request as soon as traffic to that IP address is seen. The command to remove an entry for a specific IP address is as follows:

```
sudo ip neighbor del dev <name> <IP address>
```

There is also a command that removes all entries for a specific network interface:

```
ip neighbor flush dev <name>
```

Even though the inner workings of the ARP and NDP protocols are different, all commands we discussed are applicable to both IPv4 and IPv6 neighbor tables.

Routes and route tables

IPv4 and IPv6 are routed protocols, which allows them to be used in large-scale networks that consist of multiple independent segments, such as the internet. An Ethernet network segment is flat and there are no mechanisms for grouping MAC addresses: if there are a hundred hosts in a network, the switch must maintain a MAC address table of hundred entries, and every host needs to keep the MAC address of every host it needs to communicate with in its table. That approach puts obvious limits on the maximum network size. It also makes it impossible to have multiple paths to the same part of the network, since all hosts in a data link layer segment must communicate directly.

By contrast, IP networks are broken into subnets that are connected to one another through routers— devices dedicated to forwarding packets between hosts (many routers, both home/small office and enterprise/service provider ones, are running Linux these days). The most important property of the IP addressing architecture is that subnets can be aggregated. For example, if a network internally consists of two consecutive subnets that contain 32 hosts each, say 192.0.2.0/27 and 192.0.2.32/27, then other networks can refer to it as a single network of 64 hosts—192.0.2.0/26.

Hosts and routers that are only connected to a single upstream router (typically, an internet service provider) thus can store only one route to the entire IP range: 0.0.0.0/0 for IPv4 or ::/0 for IPv6. Such a route is called a **default route**.

Viewing routing tables and routes

Let us inspect the routing table of a Linux host connected to a single router. You can view IPv4 routes with ip route show, or just ip route. Unlike ip neigh, which displays both IPv4 and IPv6 neighbors unless filtered with -4 or -6, this command defaults to IPv4 and requires the -6 option to show IPv6 routes instead:

```
$ ip route show
default via 172.20.10.1 dev eth0 proto dhcp src 172.20.10.2 metric 100
172.20.10.0/28 dev eth0 proto kernel scope link src 172.20.10.2 metric
100
```

The first entry is the default route—a route to the 0.0.0.0/0 network that covers every possible IPv4 address. The gateway is 172.20.10.1. The outgoing interface is eth0. From proto dhcp, we can infer that it was received from a DHCP server. The protocol field is purely informational, and the kernel does not use it for route selection. Internally, it is a number from 0 to 255, and some of those numbers are mapped to protocol names in the /etc/iproute2/rt_protos configuration file.

The second route to the 172.20.10.0/28 network is a connected route that simply tells the system that it can communicate with hosts from a certain subnet directly by sending packets from a certain network interface. Notice that it lacks a gateway and only has an interface field (dev). If we examine the IPv4 addresses on that machine, we will see that its address is 172.20.10.2/28:

```
$ ip address show eth0
17: eth0: <BROADCAST,MULTICAST,UP,LOWER_UP> mtu 1500 qdisc fq_codel
state UP group default qlen 1000
    link/ether 4e:79:75:4e:61:9a brd ff:ff:ff:ff:ff:ff
    altname enp0s20f0u3u3c4i2
    inet 172.20.10.2/28 brd 172.20.10.15 scope global dynamic
noprefixroute eth0
       valid_lft 65207sec preferred_lft 65207sec
    inet6 fe80::a487:4fbe:9961:ced2/64 scope link noprefixroute
       valid_lft forever preferred_lft forever
```

Whenever an IP address is added to a network interface, the kernel calculates its subnet address and adds a route to that subnet. Since the 172.20.10.0/28 subnet is smaller than 0.0.0.0/0, that route will be used for communication with hosts from that subnet rather than the default route. This is known as **the longest match rule**.

The kernel protocol number is reserved for marking such auxiliary routes that are created without a direct request from an administrator. It is possible to view routes only for a specific protocol by running the following command:

```
$ ip route show proto kernel
172.20.10.0/28 dev eth0 scope link src 172.20.10.2 metric 100
```

The kernel can also tell you which route it would use for accessing a certain host or network. For example, if you want to know how it would reach a host with a 192.0.2.1 address, run the following command:

```
$ ip route get 192.0.2.1
192.0.2.1 via 172.20.10.1 dev eth0 src 172.20.10.2 uid 1000 cache
```

Since this machine only has a single default route, the answer for every host is its default gateway — 172.20.10.1. However, on routers with multiple connections to multiple networks, the ip route get command can be helpful.

Configuring routes

Many host systems just get their default route from DHCP, but routing functionality in the Linux kernel is much more advanced.

One problem with configuring routes through the ip utility is that, as with everything configured that way, such routes only survive until the next reboot, in perfect conditions. The other problem is that if a network interface goes down (due to a disconnected cable in the case of physical network cards or protocol reset in the case of virtual links), all routes associated with that interface are permanently deleted and need to be restored by a userspace program. On enterprise and service provider routers, the userspace program is usually a routing protocol stack service such as Free Range Routing or BIRD. Those routing stack services implement dynamic routing protocols but also help manage static routes and keep them active across network interface state changes. On host systems, it can be NetworkManager or another network configuration frontend. However, it is still good to know how to create routes by hand when you need to experiment or make an emergency fix on a running machine.

To create a static route with a specific gateway address, you can use this command:

```
ip route add <network> via <gateway>
```

Let us demonstrate it on a dummy interface. First, we will create a dummy interface and assign the 192.0.2.1/24 address to it to force the kernel to create a connected route to 192.0.2.0/24 and give us space for imaginary gateways. We will use 203.0.113.0/24 for our experiments—it is another network reserved for examples and documentation that is guaranteed not to appear on the public internet, so we can be sure that we do not disturb any real traffic:

```
$ sudo ip link add name dummy1 type dummy
$ sudo ip link set dev dummy1 up
$ sudo ip address add 192.0.2.1/24 dev dummy1
```

Now we can add a route, verify that it exists, and try to take dummy1 down to see what happens:

```
$ sudo ip route add 203.0.113.0/24 via 192.0.2.10
$ ip route
default via 172.20.10.1 dev eth0 proto dhcp src 172.20.10.2 metric 100
172.20.10.0/28 dev eth0 proto kernel scope link src 172.20.10.2 metric
100
192.0.2.0/24 dev dummy1 proto kernel scope link src 192.0.2.1
203.0.113.0/24 via 192.0.2.10 dev dummy1

$ sudo ip link set dev dummy1 down
$ ip route
default via 172.20.10.1 dev eth0 proto dhcp src 172.20.10.2 metric 100
172.20.10.0/28 dev eth0 proto kernel scope link src 172.20.10.2 metric
100
```

You can see that the newly added route automatically had the right network interface in its entry: 203.0.113.0/24 via 192.0.2.10 dev dummy1. The kernel checked the route to 192.0.2.10–the address we set as a gateway—and found that it is reachable through dummy1 (nominally, at least).

When we took dummy1 down, the kernel also removed the connected route to 192.0.2.0/24 associated with it. That made the 192.0.2.10 gateway unreachable, so the kernel also removed every route whose gateway became unreachable because of that change. The kernel also does not normally allow the user to create a route whose gateway is not reachable and can detect recursive routes (that is, routes whose gateway is not reachable directly via a connected route). However, it is possible to disable that reachability check by adding a special onlink flag to the command, for example:

```
sudo ip route add 203.0.113.0/24 via 192.0.2.10 onlink
```

If you are connected to an Ethernet switch or some other kind of a multiple access network, you must specify the gateway in your routes because simply sending packets from an Ethernet device is impossible—there must be a destination address in the packet since there may be multiple hosts in the same segment. However, some network interfaces are point-to-point and have only one system on the other side. Physical point-to-point connections, such as serial WAN links, are mostly extinct now, but virtual point-to-point links are common. If you are connected to the internet via PPPoE, you can create a default route with just `sudo ip route add 0.0.0.0/0 dev ppp0` or similar—no gateway address is needed.

If you have only one route to a certain network and want to delete it, you can do it with just `sudo ip route del <network>`, and if you have multiple routes to the same network, you will need to specify the gateway or the interface to delete exactly the route you want, as in `sudo ip route del 203.0.113.0/24 via 192.0.2.10`.

There are two situations when there may be more than one route to the same destination. First, it is possible to create a backup route by setting a `metric` command for it. For example, if we create a route with a `100` metric, it will stay in the table but will not be used while routes with lower metric values still exist. If a route disappears, the kernel will automatically start using the backup route. Let us demonstrate is with the following commands:

```
$ sudo ip route add 203.0.113.0/24 via 192.0.2.10
$ sudo ip route add 203.0.113.0/24 via 192.0.2.20 metric 100
$ ip route
...
192.0.2.0/24 dev dummy1 proto kernel scope link src 192.0.2.1
203.0.113.0/24 via 192.0.2.10 dev dummy1
203.0.113.0/24 via 192.0.2.20 dev dummy1 metric 100

$ ip route get 203.0.113.100
203.0.113.100 via 192.0.2.10 dev dummy1 src 192.0.2.1 uid 1000 cache

$ sudo ip route del 203.0.113.0/24 via 192.0.2.10
$ ip route get 203.0.113.100
203.0.113.100 via 192.0.2.20 dev dummy1 src 192.0.2.1 uid 1000 cache
```

Second, the kernel can use more than one route to the same destination in parallel for load balancing and redundancy. If different paths have different bandwidths, you can specify different weights for each gateway as follows:

```
$ sudo ip route add 203.0.113.0/24 nexthop via 192.0.2.10 weight 1
nexthop via 192.0.2.20 weight 10
$ ip route
...
192.0.2.0/24 dev dummy1 proto kernel scope link src 192.0.2.1
203.0.113.0/24
    nexthop via 192.0.2.10 dev dummy1 weight 1
    nexthop via 192.0.2.20 dev dummy1 weight 10
```

Note that this method of load balancing is only applicable in networks that have direct routed connectivity throughout because packets that belong to the same connection will be sent over both paths and the return path of replies to those packets is also unpredictable. Systems that are connected to multiple ISPs but use NAT to share a public IPv4 address between all internal network hosts need more complex load-balancing configurations that ensure that entire connections are balanced and use the same outgoing path for every packet, but that is out of the scope of this book.

There are also special-purpose routes that ensure that the destination network is made unreachable. There are two types of those routes: `blackhole` and `unreachable/prohibit/throw`. Both make the kernel discard all packets sent to certain networks, but the `blackhole` route tells it to discard packets silently, while the other type also makes it send an ICMP error to the originator.

The `blackhole` route is commonly used as a crude but highly efficient filter for outgoing traffic. These routes can be used to stop hosts inside the network from communicating with a known bad destination, such as a botnet command and control node. In case of an incoming DDoS attack, they can also be used to stop its traffic at the router and keep it from reaching its target so that you can reconfigure the target host for better performance or at least avoid overloading it until the attack is over. You can create a `blackhole` route with `sudo ip route add blackhole <network>`:

```
$ sudo ip route add blackhole 203.0.113.0/24
$ ip route
...
blackhole 203.0.113.0/24
$ ping 203.0.113.10
ping: connect: Invalid argument
```

If a network is blackholed, you will not be able to connect to any host in it from the local machine. For hosts that use that machine as a router, it will look like their packets receive no replies.

The other three types of routes (`unreachable`, `discard`, and `throw`) cannot be used for DDoS protection because when the packet destination matches such a route, the kernel will not only discard the packet but also generate an ICMP packet to notify the sender that their packets are not reaching their destination, which will only make the situation worse by generating more traffic. They are best used inside corporate networks for enforcing policies in a way that will be easy to debug. If you do not want your hosts to send any traffic to a hypothetical 203.0.113.113 host, you can run `sudo ip route add prohibit 203.0.113.113/32`, and everyone who tries to connect to it will receive a message saying that the host is administratively prohibited (while with a `blackhole` route clients could not easily tell whether it is a policy or a network problem).

As you can see, the `ip` command provides rich functionality for both configuring and viewing routing and neighbor tables. Configuring routes by hand is not a common task but it is still important to know how to do it, and information retrieval commands for route and neighbor tables are very useful in day-to-day diagnostic and debugging work.

NetworkManager

Servers and embedded devices usually have fixed, statically assigned IP addresses, but desktop and especially laptop computers may need to dynamically connect to multiple networks of different types. A systems administrator with a laptop may need to connect to a wired Ethernet network in their server closet, to Wi-Fi networks in their office, home, and public spaces such as cafes, and also use a VPN tunnel to connect to the corporate network from home. Since many laptops no longer have an onboard wired network card, there may be a need to use a USB Ethernet adapter instead, so the system must handle not just on-demand network connections, but also hot-plug network devices.

Managing such configurations by hand through configuration files and commands would be tedious, so people created software projects to automate it. Those projects rely on the usual tools such as the `ip` utility and third-party projects such as strongSwan or `xl2tpd` for VPN connections but tie them under a unified user interface and include an event handling mechanism to deal with hot-plug hardware changes and users' requests to connect to different networks.

The most popular solution is the NetworkManager project, which was started by Red Hat in 2004. These days, most Linux distributions include it at least in desktop installations.

NetworkManager is a modular project. At its core, it is a daemon that handles events and keeps track of connections so that it can re-apply settings when a network interface is unplugged and then plugged back in, or the user requests a reconnect. However, most functionality for configuring the kernel and userspace tools to establish different types of connections is implemented by plugins. If you run `dnf search NetworkManager` (on Fedora or RHEL) or `apt-cache search NetworkManager` (on Debian-based distros), you will see packages with various plugins for connection types that range from well-known and widely used such as `NetworkManager-wifi` or `NetworkManager-openvpn` to obscure and experimental such as `NetworkManager-iodine`—a solution for bypassing firewalls by transmitting data inside DNS packets.

There are also multiple user interfaces for it. The network applet you can see in the tray area of desktop environment panels is a NetworkManager user interface. In MATE Desktop and many other desktop environments, you can verify that if you right-click on the network icon and choose the **About** menu point. You will see the following screen:

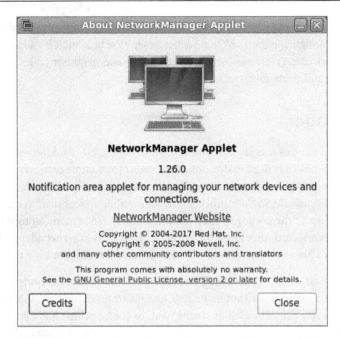

Figure 9.1 – NetworkManager applet version information dialog

In the **Edit Connections** section in the right-click menu, you can create new connections or edit existing ones:

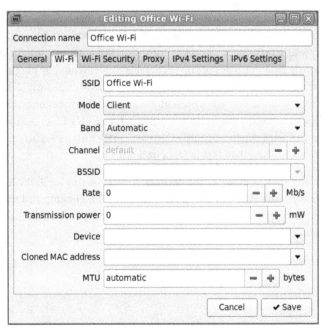

Figure 9.2 – NetworkManager connection editing dialog

Recent versions of NetworkManager save connection configuration files to a dedicated directory, while old versions would use distribution-specific formats. If you save a connection, you can find its file under /etc/NetworkManager/system-connections. Note that those files are not readable for unprivileged users. You can view the connection file for the Office Wi-Fi connection as follows:

```
$ sudo cat /etc/NetworkManager/system-connections/Office\ Wi-Fi.
nmconnection
[connection]
id=Office Wi-Fi
uuid=6ab1d913-bb4e-40dd-85a7-ae03c8b62f06
type=wifi
[wifi]
mode=infrastructure
ssid=Office Wi-Fi
[wifi-security]
key-mgmt=wpa-psk
psk=SomePassword

[ipv4]
may-fail=false
method=auto
[ipv6]
addr-gen-mode=stable-privacy
method=auto
[proxy]
```

There is also a text-mode interface for NetworkManager that can provide a GUI-like experience on headless machines. It's usually not installed by default, but on Fedora, it can be installed from the NetworkManager-tui package.

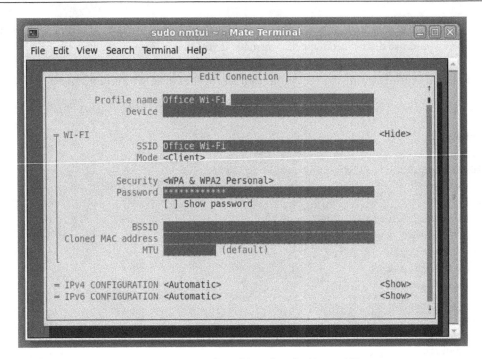

Figure 9.3 – nmtui, a text-based interface for NetworkManager

Finally, the `nmcli` tool allows managing NetworkManager connections from the command line. If you've already created a connection, you can bring it up with `nmcli connection up <name>` (similar to `nmcli connection up "Office Wi-Fi"`) or bring it down with `nmcli connection down <name>`. It also provides interactive connection editing functionality that you can use if neither `nmtui` nor a graphical desktop environment is available.

As you can see, NetworkManager makes it easy to manage typical network configurations, especially on desktop machines. Next, we will learn about distribution-specific configuration methods that do not use NetworkManager.

Distribution-specific configuration methods

NetworkManager is used by many distributions on desktop systems, but many Linux distributions have also used custom network configuration files and scripts. Some still use them, while other systems migrated to NetworkManager but maintain old methods as an alternative or in legacy long-term support releases.

Debian

In Debian, the configuration file for network interfaces is `/etc/network/interfaces`. Unlike NetworkManager's native format, this allows keeping configurations for all interfaces in a single file. To make the configuration more modular and easier to read, it is possible to store files for individual interfaces in the `/etc/network/interfaces.d/` directory.

Interface configurations are also identified by link names rather than arbitrary descriptions and UUIDs. This is how to set an onboard Ethernet device, eno1, to use static addresses for both IPv4 and IPv6, for example:

```
iface eno1 inet static
        address 203.0.113.123/24
        gateway 203.0.113.1
iface eno1 inet6 static
        address 2001:db8:abcd::123/64
        gateway 2001:db8:abcd::1
```

You can also execute custom commands when interfaces are brought up and down, using the `pre-up`, `up`, `down`, and `post-down` options. For example, to automatically add a route when eno1 goes up, run the following command:

```
iface eno1 inet static
        address 203.0.113.123/24
        gateway 203.0.113.1
        up ip route add 192.0.2.0/24 via 203.0.113.1
```

The tools for bringing interfaces up and down are named `ifup` and `ifdown`, respectively. They are only available to privileged users, so you need to run them with `sudo`, as in `sudo ifup eno1` or `sudo ifdown eno1`.

Old Red Hat-based distributions

Fedora and RHEL version 8 and above (as well as its derivatives such as Rocky Linux) use NetworkManager as their network configuration system. Up to RHEL7, however, it used a different system. Its configuration file directory was `/etc/sysconfig/network-scripts`, and each interface used its own file. For example, this is how you could statically assign the `203.0.113.113/24` address to an onboard Ethernet interface, eno1:

```
$ cat /etc/sysconfig/network-scripts/ifcfg-eno1
DEVICE=eno1
BOOTPROTO=none
ONBOOT=yes
PREFIX=24
IPADDR=203.0.113.113
```

A Red Hat-specific way to reread and apply all old-style network configurations is by using the `service network restart` command.

As you can see, distribution-specific methods are conceptually similar, although configuration syntax and names for options with the same meaning can vary wildly. If in doubt, you should always consult the documentation. Now that we've covered the most common network configuration methods, we should also learn how to verify whether the configuration works as expected.

Network troubleshooting

Network troubleshooting is a vast subject. However, most of the time, experts use the same tools that are available to every novice and those tools are not hard to learn to use at a basic level. The main difference between a novice and an expert is how well they can interpret their outputs and choose the correct options.

Using ping

The name of the `ping` utility comes from the sound of sonar—a device that uses sound pulses to discover objects underwater. That command metaphorically probes a remote host by sending an ICMP packet and listening for a reply. The sonar metaphor is a bit of a stretch because sound pulses are passively reflected, while the exchange of ICMP packets requires cooperation from a remote host.

Still, a host that runs a correctly implemented network stack should reply with an ICMP echo reply packet if it receives an echo request. At the most basic level, pinging a host tells you whether the host is online and whether there is a working network path to it.

By default, the Linux version of `ping` will keep sending echo requests indefinitely. This is in contrast with some other versions, such as that of Windows, that terminate after a finite number of packets by default. If you want the Linux `ping` to behave that way, you can specify the number of requests with `-c`, such as `-c5`, to send five requests as shown:

```
$ ping -c5 9.9.9.9
PING 9.9.9.9 (9.9.9.9) 56(84) bytes of data.
64 bytes from 9.9.9.9: icmp_seq=1 ttl=56 time=31.1 ms
64 bytes from 9.9.9.9: icmp_seq=2 ttl=56 time=24.8 ms
64 bytes from 9.9.9.9: icmp_seq=3 ttl=56 time=21.7 ms
64 bytes from 9.9.9.9: icmp_seq=4 ttl=56 time=115 ms
64 bytes from 9.9.9.9: icmp_seq=5 ttl=56 time=22.8 ms
--- 9.9.9.9 ping statistics ---
5 packets transmitted, 5 received, 0% packet loss, time 4004ms
rtt min/avg/max/mdev = 21.666/43.061/115.059/36.145 ms
```

If ICMP echo requests successfully reach their target and the target replies, then replies to those packets will come from the target address.

If packets do not reach the target host because there is no route to it, you will get replies from the last router on the path of those packets that fail to send them further because it could not find a route. For example, consider this output from pinging a host in a private network:

```
$ ping 10.217.32.33
PING 10.217.32.33 (10.217.32.33) 56(84) bytes of data.
From 10.217.41.49 icmp_seq=1 Destination Host Unreachable
From 10.217.41.49 icmp_seq=2 Destination Host Unreachable
```

In this case, the last router that received our ICMP echo request packets and failed to forward them further has a 10.217.41.49 address, so if we wanted to check why that host was inaccessible, that router would be our starting point.

Using traceroute

A ping can tell you whether a host is accessible and if not, then where exactly in the network the path of ICMP echo packets ends. However, it cannot tell you what the network path is.

To discover paths of packets through the network, you can use the traceroute utility instead. In short, that utility sends packets with an intentionally small TTL (for IPv4) or hop count (for IPv6) and records to force those routers to discard the test packet and record ICMP error messages from them to record the packet path.

Every IP packet has a field that shows how many times it was already forwarded between routers. In IPv4, that field is named **Time To Live** (**TTL**), which is a misnomer because it's not really time. In IPv6, that field is more appropriately named hop count. The purpose of those fields is the same—to prevent packets from endlessly going in circles around the network in case of a routing loop. Every time a router forwards a packet, it decrements the TTL/hop count field, and when its value reaches 0, the packet is discarded. Whenever a router discards a packet for that reason, it notifies the sender with an ICMP TTL exceeded message.

Thus, if you intentionally set the TTL of a packet to 1, it is guaranteed to be discarded by the very first router on its path. From the ICMP response about a discarded packet, you can learn the address of that router. By repeating the procedure with increasing TTL values, you can learn about every router on the path—or at least every router that cooperates and sends ICMP TTL-exceeded replies.

In the public internet and other large-scale networks, there are lots of subtleties in interpreting traceroute outputs. Some routers may not generate ICMP TTL exceeded messages at all or only generate them if their load is low, so the path may appear to have gaps (you will see * * * symbols in place of path entries where the router never generates those responses—traceroute retries its probes three times before giving up). The path shown by traceroute may not be the real path due to network segments that use **Multi-Protocol Label Switching** (**MPLS**) instead of IP routing. However, it is still a useful tool, and inside private networks, its output is usually the real path.

Here is what its output may look like:

```
$ traceroute 10.217.32.179
traceroute to 10.217.32.179 (10.217.32.179), 30 hops max, 60 byte
packets
 1   10.217.133.1 (10.217.133.1)   86.830 ms   86.769 ms   98.345 ms
 2   10.217.41.49 (10.217.41.49)   128.365 ms   128.359 ms   128.346 ms
 3   10.217.32.179 (10.217.32.179)   134.007 ms   134.024 ms   134.017 ms
```

By default, `traceroute` tries to resolve IP addresses to domain names by looking up their PTR records. It may slow it down or you may want to see raw addresses instead. If you prefer to see raw addresses, you can disable name resolution with the `-n` option.

There are many more network diagnostic tools. For example, network scanners such as *nmap* can tell you which services are running on a remote machine and gather information about its operating system and network stack. Packet capture and analysis tools such as *tcpdump* and *Wireshark* can help you get a detailed picture of network traffic. However, even with just `ping` and `traceroute`, you can verify that your network setup is working or gather debugging information to share with colleagues or support technicians.

Summary

In this chapter, we learned about the components of the Linux network stack, the tools for managing it, and different types of physical and virtual network interfaces. We learned how to assign and view addresses, view and create routes, configure network settings using NetworkManager and distribution-specific methods, and perform basic troubleshooting procedures.

In the next chapter, we will learn about managing storage devices.

10
Storage Management

It feels like there's never going to be enough space on our servers for everything that needs to be stored there. Despite the fact that the storage capacity of hard disks continues to increase and high-capacity disks are now more affordable than ever, our servers quickly fill up any available space. Our best efforts as server administrators have always been to order machines with as much storage as possible, but the reality is that even the most well-planned enterprises eventually run out of space. Additional storage space will certainly need to be added at some point in the course of administering your servers. Storage management, however, entails more than simply replacing full disks with empty ones. **Logical Volume Manager (LVM)** and other similar technologies can greatly simplify your work if you start using them as soon as possible, thus it's crucial to plan ahead.

The concepts discussed in this chapter, including LVM itself, will allow you greater freedom in the management of servers. I will also explain some other ideas that will prove useful when you work with volumes and storage on your server. In particular, the following topics will be covered:

- Creating new volumes in a filesystem
- How to format and partition storage devices
- How to mount and unmount volumes
- Learning how to use a /etc/fstab file
- Working with **Logical Volume Manager (LVM)**

Adding additional storage volumes

Additional server storage space will likely be required at some time. The capacity of a server can be expanded by installing more hard drives, either on a standalone machine or by using cloud computing. Whatever its name may be, we'll need to find out how to format and mount the device before we can use the additional space.

Using LVM (which we will cover later in this chapter), we can easily add space to an existing volume without restarting the server. However, there is an overarching procedure that must be followed when introducing a new device. The following are a few things to consider while upgrading your system's storage capacity:

- **Can you tell me how much space you'll need?** If there is enough free space in your hypervisor's storage pool, you can create a virtual disk of any size you like.

- **How do you want to call it?** Disks are typically labeled with increasing last numbers of their respective names (for example, `/dev/sdb1`, `/dev/sdb2`).

- **What format do you want for your device?** At present, ext4 is the most widely used filesystem. However, there are additional choices you might make for various tasks (such as XFS). Use ext4 if you're unsure, but research the alternatives to see if any better suit your needs. ZFS is an additional possibility; however, it is more recent than many of the other filesystems out there. You may know this already, but new Linux users may be confused by the fact that the term *filesystem* has many meanings, depending on the situation. When discussing the file and directory structure of a normal Linux system, we Linux administrators will most often use the term *filesystem*. On the other hand, the phrase can also refer to the format of a disk that is compatible with the distribution (for example, the ext4 filesystem).

- **Where do you want it mounted?** Since the new disk must be reachable by the system and possibly users, you must mount (attach) it to a directory on your filesystem from which users or applications can access it. In this chapter, we also cover LVM, and in most cases, you'll want to add it to an existing storage group. The new volume can be used with any directory you like, but I'll go over some typical ones in the *Formatting and partitioning storage devices* section. In the *Mounting and unmounting volumes* section, we will discuss the mounting and unmounting processes in detail.

Let's think about the responses to the first two queries. To determine how much more room you should implement, you should analyze your program's or company's requirements. When it comes to actual disks, your options are limited to which disk to buy. With virtual drives, you can be more thrifty by adding a smaller disk to serve your needs (you can always add more later). The primary advantage of using LVM with virtual drives is the ability to increase a filesystem without restarting the server. If you have a 50 GB volume and want to make it larger, you can create two more 10 GB virtual disks and enlarge it that way.

While LVM is not limited to virtual servers, it can be used on physical servers as well; however, doing so would likely necessitate a reboot due to the need to open the casing and physically attach a hard drive. The ability to add or remove physical hard disks from some servers is called **hot-plugging**, and it's a huge time-saver that doesn't need turning the server off. Finally, you may get the device's name by typing `fdisk -l` into a terminal. We can find out what our new disk is called by using the `fdisk` command, which is more commonly used to create and erase partitions. If you run the `fdisk -l` command as `root` or with `sudo`, you'll get the following details:

```
sudo fdisk -l
```

```
[voxsteel@hp-centos8 ~]$ sudo fdisk -l
[sudo] password for voxsteel:
Disk /dev/sda: 119.2 GiB, 128035676160 bytes, 250069680 sectors
Units: sectors of 1 * 512 = 512 bytes
Sector size (logical/physical): 512 bytes / 4096 bytes
I/O size (minimum/optimal): 4096 bytes / 4096 bytes
Disklabel type: dos
Disk identifier: 0xc2c39d4f

Device     Boot    Start      End   Sectors    Size Id Type
/dev/sda1   *        2048  2099199   2097152     1G 83 Linux
/dev/sda2         2099200 250068991 247969792 118.2G 8e Linux LVM

Disk /dev/mapper/cl-root: 70 GiB, 75161927680 bytes, 146800640 sectors
Units: sectors of 1 * 512 = 512 bytes
Sector size (logical/physical): 512 bytes / 4096 bytes
I/O size (minimum/optimal): 4096 bytes / 4096 bytes

Disk /dev/mapper/cl-swap: 3.9 GiB, 4131389440 bytes, 8069120 sectors
Units: sectors of 1 * 512 = 512 bytes
Sector size (logical/physical): 512 bytes / 4096 bytes
I/O size (minimum/optimal): 4096 bytes / 4096 bytes

Disk /dev/mapper/cl-home: 44.4 GiB, 47664070656 bytes, 93093888 sectors
Units: sectors of 1 * 512 = 512 bytes
Sector size (logical/physical): 512 bytes / 4096 bytes
I/O size (minimum/optimal): 4096 bytes / 4096 bytes
[voxsteel@hp-centos8 ~]$
```

Figure 10.1 – Listing all disks

Now that we have learned how to list all the disks, we will next learn how to format and partition storage devices.

Formatting and partitioning storage devices

A disk must be formatted before it can be used. Finding the device's given name is necessary to format the right disk. If you read the last section, you know that drives in Linux distributions follow a predetermined naming convention. Consequently, you should be familiar with the new disk's device name. You can view information about the storage devices connected to your server using the sudo fdisk -l command, as previously mentioned:

```
sudo fdisk -l
```

```
[voxsteel@hp-centos8 ~]$ sudo fdisk -l
Disk /dev/sda: 119.2 GiB, 128035676160 bytes, 250069680 sectors
Units: sectors of 1 * 512 = 512 bytes
Sector size (logical/physical): 512 bytes / 4096 bytes
I/O size (minimum/optimal): 4096 bytes / 4096 bytes
Disklabel type: dos
Disk identifier: 0xc2c39d4f

Device     Boot    Start        End    Sectors   Size Id Type
/dev/sda1  *        2048    2099199    2097152    1G 83 Linux
/dev/sda2        2099200  250068991  247969792 118.2G 8e Linux LVM


Disk /dev/mapper/cl-root: 70 GiB, 75161927680 bytes, 146800640 sectors
Units: sectors of 1 * 512 = 512 bytes
Sector size (logical/physical): 512 bytes / 4096 bytes
I/O size (minimum/optimal): 4096 bytes / 4096 bytes


Disk /dev/mapper/cl-swap: 3.9 GiB, 4131389440 bytes, 8069120 sectors
Units: sectors of 1 * 512 = 512 bytes
Sector size (logical/physical): 512 bytes / 4096 bytes
I/O size (minimum/optimal): 4096 bytes / 4096 bytes


Disk /dev/mapper/cl-home: 44.4 GiB, 47664070656 bytes, 93093888 sectors
Units: sectors of 1 * 512 = 512 bytes
Sector size (logical/physical): 512 bytes / 4096 bytes
I/O size (minimum/optimal): 4096 bytes / 4096 bytes


Disk /dev/sdb: 7.5 GiB, 8004304896 bytes, 15633408 sectors
Units: sectors of 1 * 512 = 512 bytes
Sector size (logical/physical): 512 bytes / 512 bytes
I/O size (minimum/optimal): 512 bytes / 512 bytes
Disklabel type: gpt
Disk identifier: 1626D7B5-0137-0648-8AE7-774CD98BA1DB
```

Figure 10.2 – Listing all disks

The /dev/sdb device is new to my server, as I just installed it (see *Figure 10.2*). I'm using it for the exercises in this chapter. Currently, it is not partitioned. At this point, it's quite clear that the storage device referenced by /dev/sdb in the previous example is brand new. To avoid losing data, we must be careful never to format or repartition the wrong device. Since there are no partitions on /dev/sdb (as this volume wasn't there before I added it), it's evident that this is the disk we should work with. One or more partitions can be made on it at this point, bringing us one step closer to actually using it.

Using the fdisk command with sudo and the device's name as an option, we can partition the drive. The following is the command I would use to access disk /dev/sdb:

```
sudo fdisk /dev/sdb
```

You'll see that I didn't specify a partition number because `fdisk` deals with the disk directly (and we also have yet to create any partitions). In this section, I will assume that you have access to a drive that hasn't been partitioned yet or is completely erasable. Upon successful execution, `fdisk` will display an introductory message and prompt.

To see a list of available commands, use the m key on your keyboard, as shown in the following screenshot:

```
Command (m for help): m

Help:

  DOS (MBR)
   a   toggle a bootable flag
   b   edit nested BSD disklabel
   c   toggle the dos compatibility flag

  Generic
   d   delete a partition
   F   list free unpartitioned space
   l   list known partition types
   n   add a new partition
   p   print the partition table
   t   change a partition type
   v   verify the partition table
   i   print information about a partition

  Misc
   m   print this menu
   u   change display/entry units
   x   extra functionality (experts only)

  Script
   I   load disk layout from sfdisk script file
   O   dump disk layout to sfdisk script file

  Save & Exit
   w   write table to disk and exit
   q   quit without saving changes

  Create a new label
   g   create a new empty GPT partition table
   G   create a new empty SGI (IRIX) partition table
   o   create a new empty DOS partition table
   s   create a new empty Sun partition table

Command (m for help): []
```

Figure 10.3 – Help menu for fdisk

Here, I will demonstrate the sequence of steps necessary to create a new disk. Understand that `fdisk` can do serious damage to your system. Using `fdisk` on the wrong drive can cause permanent data loss. It's normal for administrators to internalize the use of tools such as `fdisk` to the point where it's second nature; however, before executing any such instructions, double-check that you're indeed accessing the correct disk. There are two types of partition tables – the **Master Boot Record** (**MBR**) and the **GUID Partition Table** (**GPT**) – that need to be discussed before we can move further with creating a new partition. You can choose between the MBR and GPT partition tables when setting up a new disk's partition table. If you've worked with servers for a while, you've undoubtedly used MBR, while GPT is the current standard. On older computers, the MBR partition structure can be referred to as the DOS partition structure. You may be aware that **DOS** refers to the **Disk Operating System**; nevertheless, in this chapter, we will refer to the partitioning scheme developed by IBM many years ago. For instance, the MBR partition structure is referred to as *DOS* while using `fdisk`. There are restrictions to think about when using MBR partition tables. To begin with, there is a limit of four primary partitions in MBR. It also restricts your disk use to about 2 TB. As long as your disk is under 2 TB in size, you should be fine, but disks greater than 2 TB are becoming increasingly prevalent. However, GPT doesn't limit partition sizes, so if you have a disk that's many TB in size, the choice between MBR and GPT is already made for you. In addition, `fdisk` with a GPT partition table allows you to construct up to 128 primary partitions. The following command gives options on what to do with the specific HDD, `/dev/sdb`:

```
[voxsteel@hp-centos8 ~]$ sudo fdisk /dev/sdb

Welcome to fdisk (util-linux 2.32.1).
Changes will remain in memory only, until you decide to write them.
Be careful before using the write command.


Command (m for help): n
All space for primary partitions is in use.

Command (m for help): g
Created a new GPT disklabel (GUID: 1626D7B5-0137-0648-8AE7-774CD98BA1DB).
The old dos signature will be removed by a write command.

Command (m for help): n
Partition number (1-128, default 1): 2
First sector (2048-15633374, default 2048):
Last sector, +sectors or +size{K,M,G,T,P} (2048-15633374, default 15633374): +2G

Created a new partition 2 of type 'Linux filesystem' and of size 2 GiB.

Command (m for help):
```

Figure 10.4 – Creating a new partition

I used GPT, so I had to enter g:

```
m    print this menu
u    change display/entry units
x    extra functionality (experts only)

Script
I    load disk layout from sfdisk script file
O    dump disk layout to sfdisk script file

Save & Exit
w    write table to disk and exit
q    quit without saving changes

Create a new label
g    create a new empty GPT partition table
G    create a new empty SGI (IRIX) partition table
o    create a new empty DOS partition table
s    create a new empty Sun partition table


Command (m for help): g
Created a new GPT disklabel (GUID: 49CC6AB5-C5C6-4A49-B0BC-864A11A4467B).
The old dos signature will be removed by a write command.
```

Figure 10.5 – Creating a new GPT partition table

Don't forget to enter w to save changes.

If you made any mistakes, you can run fdisk again. After entering the fdisk prompt again, you can construct a new GPT layout by hitting g, or a new MBR layout by typing o if you made a mistake or simply wish to start over. Partitioning your drive is a two-step process, so you'll need to repeat the previous stages. You can try this out for yourself a few times until you get the hang of it.

Now, using fdisk -l /dev/sdb, we can see there is a new partition, /dev/sdb2, as shown in the following screenshot:

```
[voxsteel@hp-centos8 ~]$ sudo fdisk -l /dev/sdb
Disk /dev/sdb: 7.5 GiB, 8004304896 bytes, 15633408 sectors
Units: sectors of 1 * 512 = 512 bytes
Sector size (logical/physical): 512 bytes / 512 bytes
I/O size (minimum/optimal): 512 bytes / 512 bytes
Disklabel type: gpt
Disk identifier: 1626D7B5-0137-0648-8AE7-774CD98BA1DB

Device      Start     End Sectors Size Type
/dev/sdb2    2048 4196351 4194304   2G Linux filesystem
[voxsteel@hp-centos8 ~]$
```

Figure 10.6 – Listing the partitions for /dev/sdb HDD

Now that we have learned how to create a new partition, we will see how to format it.

Formatting a newly created partition

Your new partition can be formatted once you've finished designing the disk's partition arrangement and are satisfied with it. The results of sudo fdisk -l will be different now that I have partitioned the new drive.

A new partition, /dev/sdb2, has been created and is reflected in the output. We can proceed with the formatting at this time. The mkfs command is used to create the filesystem. In order to execute this operation, you must use the correct syntax, which consists of entering mkfs, followed by a period (.), and then the name of the filesystem you wish to create. Using this code as an example, we can format /dev/sdb2 as ext4 by running the sudo mkfs.ext4 /dev/sdb2 command:

```
[voxsteel@hp-centos8 ~]$ sudo mkfs.ext4 /dev/sdb2
mke2fs 1.45.6 (20-Mar-2020)
Creating filesystem with 524288 4k blocks and 131072 inodes
Filesystem UUID: 75019d8b-ff26-490a-821c-2754239369af
Superblock backups stored on blocks:
        32768, 98304, 163840, 229376, 294912

Allocating group tables: done
Writing inode tables: done
Creating journal (16384 blocks): done
Writing superblocks and filesystem accounting information: done
```

Figure 10.7 – Formatting a partition

It is important to remember to format the partition; otherwise, it won't be usable.

Mounting and unmounting volumes

The next step after adding and formatting a new storage volume on your server is mounting the device. The mount command accomplishes this task. With this command, you can link a removable drive (or a network share) to a directory on the server's hard drive. Mounting requires a clean directory. In order to mount a device, you must specify a directory to mount it to by using the mount command, which we will practice with an example shortly. Mounting additional storage is as simple as issuing the mount command and selecting a location that isn't currently mounted or full of data. mount is a command that normally requires root privileges to execute. However, in most cases, only the root should mount volumes (although there is a workaround that involves allowing regular users to mount volumes; we won't discuss that right now). Since you need a directory in which to mount these volumes, I'll show you how to make one called /usbpartition using the following command:

```
sudo mkdir /usbpartition
sudo mount /dev/sdb2 /usbpartition
```

With the preceding command, I mount the /dev/sdb2 device to the /usbpartition directory as an example. Obviously, you'll need to change the /dev/sdb2 and /usbpartition references to reflect your own device and directory choices. A mount normally requires the -t option to specify the device type, but fdisk -l can be used as a handy reminder if you've forgotten what devices are installed on your server. The mount command that I should have used with the -t option, given that my disk is ext4-formatted, is as follows:

```
sudo mount /dev/sdb2 -t ext4 /usbpartition
```

To check whether this has been mounted properly, use this command:

```
mount -v | grep usbpartition
```

If you're done working with a volume, you can unmount it with the umount command (the *n* in *unmount* is left out on purpose):

```
umount /usbpartition
```

It is possible to remove a storage device from your filesystem using the umount command, which also requires you to be logged in as root or with sudo. The volume must be turned off for this command to take effect. A device or resource busy error message may appear if this is the case. After unmounting, you can verify that the filesystem is no longer mounted by running df -h and noting that it does not return any results. When devices are manually mounted, they will unmount when the server reboots. I will show you how to update the /etc/fstab file so that the mount is available at server startup in the following section.

Updating the /etc/fstab file

An essential Linux system file is the /etc/fstab file. You can manually mount additional volumes at boot time by editing this file. However, this file's primary function is to mount your primary filesystem, thus any mistakes you make while modifying it will prevent your server from starting up (at all). Take extreme caution.

The location of the root filesystem is determined by reading the /etc/fstab file, which is read during system boot. This file is also used to determine the swap partition's location, and it is mounted at boot time. Each mount point in this file will be read and mounted sequentially by your system. This file can be used to automatically mount almost any type of storage device. You can even install Windows server network shares. In other words, it has no morals and won't pass judgment (unless you make a typo).

This is what an /etc/fstab file looks like:

```
[[voxsteel@hp-centos8 ~]$ more /etc/fstab

#
# /etc/fstab
# Created by anaconda on Wed Oct 26 13:58:55 2022
#
# Accessible filesystems, by reference, are maintained under '/dev/disk/'.
# See man pages fstab(5), findfs(8), mount(8) and/or blkid(8) for more info.
#
# After editing this file, run 'systemctl daemon-reload' to update systemd
# units generated from this file.
#
/dev/mapper/cl-root           /                    xfs     defaults       0 0
UUID=7d6941ee-d099-4035-8bdc-509527e6a3b8 /boot               xfs      defaults       0 0
/dev/mapper/cl-home     /home              xfs     defaults       0 0
/dev/mapper/cl-swap     none               swap    defaults       0 0
```

Figure 10.8 – A sample of an /etc/fstab file

It is important to be careful when editing this file, as incorrect settings can cause the system to fail to boot or cause data loss. It is recommended to make a backup of the file before making any changes.

Editing /etc/fstab file

As we mentioned earlier, any device mounted manually won't be mounted automatically on reboot.

In order to do it automatically, the device has to be added to the /etc/fstab file.

I added an entry in /etc/fstab to mount /dev/sdb2 automatically in /usbpartition/ on boot, as shown in the last line of the following screenshot:

```
[[voxsteel@hp-centos8 ~]$ more /etc/fstab

#
# /etc/fstab
# Created by anaconda on Wed Oct 26 13:58:55 2022
#
# Accessible filesystems, by reference, are maintained under '/dev/disk/'.
# See man pages fstab(5), findfs(8), mount(8) and/or blkid(8) for more info.
#
# After editing this file, run 'systemctl daemon-reload' to update systemd
# units generated from this file.
#
/dev/mapper/cl-root           /                    xfs     defaults       0 0
UUID=7d6941ee-d099-4035-8bdc-509527e6a3b8 /boot               xfs      defaults       0 0
/dev/mapper/cl-home     /home              xfs     defaults       0 0
/dev/mapper/cl-swap     none               swap    defaults       0 0
/dev/sdb2               /usbpartition/     ext4    defaults       0 0
```

Figure 10.9 – A sample of the /etc/fstab file

On my machine, both the fifth and sixth columns read 0, indicating a successful dump and a passing grade. Nearly invariably set to 0, the dump partition can be checked by a backup program to see whether the filesystem needs to be backed up (0 for no, and 1 for yes). Since almost nothing uses this anymore, you can usually just leave it at 0. Filesystems will be checked in the order specified in the pass field. In the event of a system crash or as part of a routine maintenance procedure, the `fsck` utility checks drives for filesystem problems. Values can be either a 0 or a 1. When set to 0, `fsck` will never run to check on the partition. If this value is 1, the partition is examined before anything else.

By default, only the root user can modify the `/etc/fstab` file, so it's important to be careful when editing it. Incorrectly modifying this file can cause serious problems with the system's boot process and data integrity.

Utilizing LVM

Your organization's requirements will evolve over time. As server administrators, we constantly strive to set up resources with future expansion in mind. Unfortunately, budgets and policy shifts frequently get in the way. You'll find that LVM is invaluable in the long run. Linux's superiority in scalability and cloud deployments is due in large part to technologies such as LVM. By using LVM, you can expand or contract your filesystems without having to restart the server. Consider the following scenario. Say you have a mission-critical app operating on a virtualized production server. It's possible that when you initially set up the server, you allocated 300 GB of space for the application's data directory, thinking it would never grow to fill that much space. The expansion of your company has not only increased your space requirements but also created a crisis. So, tell me, what do you do? If the server was initially configured to use LVM, then adding a new storage volume, including it in the LVM pool, and expanding the partition would not require a reboot. However, without LVM, you'd have to schedule downtime for your server while you added more storage the old-fashioned way, which could take hours. Even if your server isn't virtual, you can still profit from expanding your filesystem online by installing more hard drives and leaving them on standby without using them. In addition, you can add more volumes without having to shut down the server if your server supports hot-plugging. Because of this, I can't stress enough how important it is to use LVM whenever possible on storage volumes in virtual servers. Again, LVM is essential to set up storage volumes on a virtual server. If you don't, you'll have to spend your weekends laboring to add disks, since you will have ran out of space during the week.

Getting started with LVM

Linux's server installer allows you to select LVM as a volume manager for a fresh server setup. However, LVM should be utilized extensively for storage volumes, especially those that will house user and application data. If you want the root filesystem on your Ubuntu server to take advantage of LVM's functionality, LVM is a good option. We'll need to have a firm grasp on volume groups, physical volumes, and logical volumes before we can begin using LVM. The logical and physical volumes you intend to use with an LVM solution are organized into volume groups. In a nutshell, a volume group is the umbrella term for your whole LVM infrastructure. You can think of it as a container that can hold disks. A `vg-accounting` volume group is an illustration of this type of group. The accounting department would use this volume group to store their records. It will include both the actual disk space and the virtual disk space that these users will access. It's worth noting that you're not restricted to a single volume group but can instead create multiple, each with its own set of disks and volumes. A physical volume is a disk, either real or virtual, that belongs to a volume group. A vg-accounting volume group might theoretically have three 500 GB hard drives, each of which would be treated as a physical volume. Keep in mind that, although these disks are virtual, in the context of LVM they are still referred to as physical volumes. A physical volume is a storage device that belongs to a volume group. Finally, the concept of logical volumes is comparable to that of partitions. In contrast to traditional partitions, logical volumes can span over numerous physical drives. A logical volume, for instance, may be set up with three 500 GB disks, giving you access to a total of 1,500 GB. When it's mounted, it acts like a single partition on a regular hard drive, allowing users to save and access files with the same ease as they would with any other disk. When the space on the volume runs out, you can expand it by adding a new disk and then expanding the partition. Although it may actually be comprised of several hard drives, to your users it will appear as a single, unified space. Physical volumes can be subdivided in any way that makes sense to you. Users will interact with logical volumes, which are created from the underlying physical volumes.

Installing LVM on a server that isn't already utilizing it requires having at least one unused volume and the necessary packages, either of which may or may not already be present on the server. The following commands will tell you whether the necessary lvm2 package is present on your server:

For Debian, use this command:

```
sudo apt search lvm2 | grep installed
```

For Redhat, use this command:

```
sudo yum list installed|grep lvm2
```

The next step is to count all of the disks in our possession. Many times now, we have used the fdisk -l command to display a list of these drives. As an example, I have /dev/sdc on my server now. Disk names will vary by hardware or virtualization platform, so you'll need to tweak the following commands to work with your setup. First, we must prepare each disk for usage with LVM by creating a physical volume. It is important to remember that setting up LVM does not include formatting a storage device or using fdisk to configure it. In this case, formatting occurs later. To begin setting up our drives for usage with LVM, we will use the pvcreate command. As a result, we must execute the pvcreate command on each of the drives we intend to employ. To set up LVM with my USB disks, I will execute the following:

```
sudo pvcreate /dev/sdc
```

If you want to check that everything is set up properly, you can see a list of your server's physical volumes by running the `pvdisplay` command as root or using `sudo`:

```
"/dev/sdc" is a new physical volume of "7.45 GiB"
--- NEW Physical volume ---
PV Name                 /dev/sdc
VG Name
PV Size                 7.45 GiB
Allocatable             NO
PE Size                 0
Total PE                0
Free PE                 0
Allocated PE            0
PV UUID                 LwiR7P-BfEg-mOyY-ywec-OyBG-R7Ue-iZ5VAP
```

Figure 10.10 – The pvdisplay command output

Only one volume could be displayed on this page, so only that one is shown in the screenshot. If you scroll up, more output from the `pvdisplay` command will be displayed. We have access to a number of physical volumes, but none of them have been placed in a volume group. Actually, we haven't done something as simple as making a volume group yet. Using the `vgcreate` command, we can create a volume group, give it a name, and add our first disk to it:

```
sudo vgcreate vg-packt /dev/sdc
```

At this point, I creating a volume group called `vg-packt` and allocate one of the created physical volumes (`/dev/sdc`) to it. With the `sudo vgdisplay` command, we can see the volume group's configuration, including the number of disks it uses (which should be 1 at this point):

Right now, all that has to be done is for us to make a logical volume and format it. The disk space we allocate to our volume group may be used in its entirety, or in part. Here's the command I'll use to partition the newly added virtual disk within the volume group into a 5 GB logical volume:

```
sudo lvcreate -n packtvol1 -L 5g vg-packt
```

Although the command looks difficult, it is actually quite simple. For clarity, I use the `-n` option to give my logical volume the name `packtvol1` in this example. I use the `-L` option followed by `5g` to specify that I only want to allocate 5 GB of space. The volume group that this logical volume will be part of is listed as the final item. To view details about this volume, use the `sudo lvdisplay` command:

```
~ $sudo lvdisplay
  --- Logical volume ---
  LV Path                /dev/vg-packt/packtvol1
  LV Name                packtvol1
  VG Name                vg-packt
  LV UUID                qQ3zmr-e1KQ-e2UJ-0Ksr-9P5d-HmAv-NAJvUX
  LV Write Access        read/write
  LV Creation host, time hp-centos8, 2022-11-06 17:29:14 +0000
  LV Status              available
  # open                 0
  LV Size                5.00 GiB
  Current LE             1280
  Segments               1
  Allocation             inherit
  Read ahead sectors     auto
  - currently set to     8192
  Block device           253:3
```

Figure 10.11 – The lvdisplay command output

At this point, we have everything we need to set up LVM. Like with non-LVM disks, we must format a volume before using it.

Creating a format for logical disks

The next step is to utilize the correct format for our logical volume. However, for the formatting process to go smoothly, we must always know the device's name. Since LVM exists, this is a breeze. You can see this in the output (it's the third line down in *Figure 10.11*, under **LV Path**) that the lvdisplay command provided. Let's use the ext4 filesystem to set up the drive:

```
sudo mkfs.ext4 /dev/vg-packt/packtvol1
```

Finally, this storage device can be used like any other hard drive. Mine will be mounted at /mnt/lvm/packtvol1, but you can use whatever you like:

```
sudo mount /dev/vg-packt/packtvol1 /mnt/lvm/packtvol1
```

We can run df -h to verify that the volume is mounted and displays the right size. The single disk in our current LVM arrangement makes this useless. The 5 GB I've allotted probably won't last very long, but we have some unused space that we can put to good use:

```
~$ df -h | grep packtvol1
/dev/mapper/vg--packt-packtvol1   4.9GB   20M.   4.6GB   1%   /mnt/lvm/
packtvol1
```

The following `lvextend` command allows me to expand my logical volume to fill the remaining space on the physical disk:

```
sudo lvextend -n /dev/vg-packt/packtvol1 -l +100%FREE
```

```
~ $sudo lvextend -n /dev/vg-packt/packtvol1 -l +100%FREE
  Size of logical volume vg-packt/packtvol1 changed from 5.00 GiB (1280 extents) to 7.45 GiB (1908 extents).
  Logical volume vg-packt/packtvol1 successfully resized.
~ $
```

Figure 10.12 – The lvextend command

To be more specific, the preceding `+100%FREE` option specifies that we wish to allocate the full remaining space to the logical volume. In my case, this is only 2.5 GB, as I used a USB stick for demo purposes.

All of the space on the physical drive I designated for my logical volume is being used up. However, tread carefully, because if I had more than one physical volume allocated, the command would have taken up all of that space as well, making the logical volume the size of all the space on all the disks. Even if you don't want to do this all the time, it's fine by me because I just have one physical volume. Feel free to use the `df  -h` tool once again to verify your available storage space:

```
~$ df -h | grep packtvol1
/dev/mapper/vg--packt-packtvol1   4.9GB   20M.   4.6GB   1%   /mnt/lvm/
packtvol1
```

Unfortunately, the additional volume space we've added isn't reflected. `df` still returns the old volume size in its output. The reason for this is that we did not resize the `ext4` filesystem that is located on this logical disk, despite the fact that we have a larger logical volume and it has all the space given to it. The `resize2fs` command is what we'll utilize to do this:

```
sudo resize2fs /dev/mapper/vg--packt-packtvol1
```

```
~ $sudo resize2fs /dev/mapper/vg--packt-packtvol1
[sudo] password for voxsteel:
resize2fs 1.45.6 (20-Mar-2020)
Filesystem at /dev/mapper/vg--packt-packtvol1 is mounted on /mnt/lvm/packtvol1; on-line resizing required
old_desc_blocks = 1, new_desc_blocks = 1
The filesystem on /dev/mapper/vg--packt-packtvol1 is now 1953792 (4k) blocks long.
```

Figure 10.13 – The resize2fs command

Now, using the `df  -h` command again, we can see we have all the space allocated:

```
~$ df -h | grep packtvol1
/dev/mapper/vg--packt-packtvol1   7.3GB   23M.   6.9GB   1%   /mnt/lvm/
packtvol1
```

In this section, we've learned how to use logical volumes and how to extend a filesystem.

Deleting volumes with LVM

Last but not least, you probably want to know what happens when you delete a logical volume or volume group. The lvremove and vgremove commands are used for this reason. Destructive as they may be, these commands can be very helpful if you ever need to get rid of a logical volume or volume group. The following syntax will get rid of any logical volumes:

```
sudo lvremove vg-packt/packtvol1
```

Giving the lvremove command the name of the volume group you want to remove the logical volume from, followed by a forward slash, is all that's required. To remove the entire volume group, the following command should do the trick:

```
sudo vgremove vg-packt
```

Even though you probably won't be removing logical volumes very often, there are commands available to help you do so if you ever find yourself in need of decommissioning an LVM component. Hopefully, you can see why LVM is so great now.

The pvremove command in Linux is used to remove a **physical volume** (**PV**) from the LVM. Before using this command, make sure that the PV you want to remove is not part of any volume group and does not contain any active logical volumes. Otherwise, data loss may occur.

This technology gives you unprecedented control over the data stored on your server. Linux's versatility in the cloud is due in part to LVM's adaptability. If you're not familiar with LVM, these ideas may seem foreign at first. However, with virtualization, experimenting with LVM is straightforward. Until you feel comfortable making, editing, and erasing volume groups and logical volumes, I advise you to put in some practice time. Concepts that aren't immediately evident will become so with repeated exposure.

Summary

Maintaining a smooth operation requires careful storage management, as a full filesystem will cause your server to cease. Fortunately, Linux servers come with a plethora of storage management capabilities, some of which are the envy of competing systems. It would not be possible to perform our jobs as Linux server administrators without innovations such as LVM. In this chapter, we dove into these resources and learned some storage management tricks. We went through a wide range of topics, including how to create and manage partitions, mount and unmount volumes, work with the fstab file, use LVM, and check disk use.

In the next chapter, we will discuss logging configuration and remote logging.

Part 3: Linux as a Part of a Larger System

All modern IT infrastructures consist of multiple machines with different roles, so all systems administrators need to know how to make their Linux-based systems work together. In this part of the book, you will learn how to collect log messages from all these systems on a central server, simplify user account and permission management by using centralized authentication mechanisms, create robust services with automatic failover and load balancing, and manage multiple systems at once with automation tools. You will also learn how to keep your systems secure.

This part has the following chapters:

- *Chapter 11, Logging Configuration and Remote Logging*
- *Chapter 12, Centralized Authentication*
- *Chapter 13, High Availability*
- *Chapter 14, Automation with Chef*
- *Chapter 15, Security Guidelines and Best Practices*

11
Logging Configuration and Remote Logging

Logging is an important aspect of any operating system, including Linux. It provides a way to collect and analyze system events and activities, which can be useful for troubleshooting, monitoring, and auditing purposes. In this chapter, we will explore the different aspects of logging configuration and remote logging in Linux.

In this chapter, we will cover the following topics:

- Logging configuration
- Log rotation
- Journald
- Log forwarding

Logging configuration

Linux uses the syslog system for logging. The syslog daemon collects messages from different parts of the system and writes them to log files. The syslog configuration file is usually located at `/etc/syslog.conf` or `/etc/rsyslog.conf`, depending on the distribution. This file contains the rules that specify which messages to log and where to store them.

There is a critical parameter called `facility.severity` that is a crucial part of the logging configuration in Linux. It allows you to control which log messages should be recorded and where they should be stored. The facility and severity can be specified either numerically or using their symbolic names. For example, the following rule logs all messages with a severity level of warning or higher from the auth facility to the `/var/log/auth.log` file:

```
auth.warning /var/log/auth.log
```

The target part of the configuration file specifies where to store the logged messages. The target can be a file, a remote host, or a program that processes the messages. The target syntax is as follows:

```
target_type(target_options)
```

The target type can be one of the following:

- `file`: Stores the messages in a local file
- `remote`: Sends the messages to a remote host using the syslog protocol
- `program`: Sends the messages to a local program for processing

For example, the following rule sends all messages with a severity level of error or higher to a remote host with an IP address of `192.168.1.100` using the syslog protocol:

```
*.err @192.168.1.100
```

After modifying the syslog configuration file, the syslog daemon must be restarted to apply the changes. The command to restart the syslog daemon varies depending on the distribution. For example, on Ubuntu, the command is as follows:

```
sudo service rsyslog restart
```

A log message's primary focus is on log data. Alternatively stated, log data is the explanation behind a log message. If you use an image, file, or other resources on a website, the server that hosts your site will likely keep track of that fact. You may see who accessed a certain resource by examining the log data – in this case, the user's username.

The term *logs* is shorthand for a collection of log messages that can be pieced together to provide context for an event.

Each entry in the log file can be roughly classified as one of the following:

- **Informational**: Purposely vague, these messages aim to inform users and administrators that a positive change has occurred. For instance, Cisco IOS will notify appropriate parties whenever the system reboots. However, caution is required. If a restart occurs at an inconvenient time, such as outside of maintenance or business hours, you may have cause for concern. The next few chapters of this book will teach you the knowledge and techniques you'll need to deal with a situation like this.

- **Debug**: When something goes wrong with running application code, debug messages are sent by the system to help developers identify and address the problem.

- **Warning**: This is issued when something is lacking or needed for a system, but not to the point where its absence would prevent the system from functioning. Some programs may log a notice to the user or operator if they don't receive the expected number of arguments on the command line, even though they can still operate normally without them.

- **Error**: In the event of an error, the computer system stores the relevant information in a log that may be analyzed later. An OS might generate an error log, for instance, if it is unable to synchronize buffers to disk. Unfortunately, many error messages simply provide a broad outline of the problem. More investigation is usually required to determine the root of a problem.

- **Alert**: The purpose of an alert is to draw your attention to a noteworthy development. In most cases, notifications will come from security-related devices and systems, although this is not always the case. All incoming data to a network can be inspected by an **intrusion prevention system (IPS)** placed at its entrance. It examines the information included in the packets to determine whether or not to enable a certain network connection. The IPS can react in several predetermined ways if it detects a potentially malicious connection. The action and the decision will be documented.

Next, we'll quickly go through the processes involved in transmitting and collecting log data. Then, we'll discuss what a log message is.

How does log data get sent and gathered?

It's easy to send and gather log data. Syslog is a protocol used for sending and gathering log data in computer networks. It is a standard protocol that allows different devices to send log messages to a central logging server or device.

Here's how it typically works:

1. A device generates a log message. This could be a server, network device, application, or any other device that generates logs.

2. The device sends the log message to a syslog server using the syslog protocol. The syslog server can be located on-premises or in the cloud.

3. The syslog server receives the log message and processes it. It can store the log message in a file or database, or forward it to other systems for further analysis.

4. The syslog server can also apply filters and rules to the log messages it receives. For example, it can discard log messages that are not relevant or send an alert when a critical error occurs.

5. System administrators and analysts can access the log data stored in the syslog server for troubleshooting, analysis, and reporting.

The following are some of the advantages of using a centralized log collector:

- It's a centralized repository for all of your log messages

- Logs can be stored there for safekeeping

- This is where all of your server's log information may be inspected

Log analysis is crucial to the health of applications and server architecture, but it can be laborious if data is dispersed in multiple locations. Why not have just one consolidated logbook rather than a bunch of individual ones? Rsyslog may be the solution you've been looking for.

Using a centralized logging system, you may collect the logs from all of your servers and programs into one centralized area. In addition, this tutorial will help you implement centralized logging on Linux nodes by use of the rsyslog configuration.

This section is meant to be a practical example.

Checking rsyslog service on all servers

A high-performance log processing system called rsyslog is pre-installed and ready to use on both Debian and RHEL systems.

The syslog protocol has been improved with rsyslog, which gives it more contemporary and dependable features. These additional features include a large number of inputs and outputs, a modular design, and excellent filtering.

The most recent version of rsyslog as of this writing is v8.2112.0. Therefore, you will verify the rsyslog service's status and the version of rsyslog installed on your computer. This will guarantee that rsyslog is running at its most recent version.

Open a command prompt and use the following sudo su command to take control of all servers. When prompted, enter your password.

In the following screenshot, you'll find that Centos 8 ships with rsyslog v8.2102.0 by default:

```
[[voxsteel@centos8 rsyslog.d]$ rsyslogd -version
rsyslogd  8.2102.0-5.el8 (aka 2021.02) compiled with:
        PLATFORM:                               x86_64-redhat-linux-gnu
        PLATFORM (lsb_release -d):
        FEATURE_REGEXP:                         Yes
        GSSAPI Kerberos 5 support:              Yes
        FEATURE_DEBUG (debug build, slow code): No
        32bit Atomic operations supported:      Yes
        64bit Atomic operations supported:      Yes
        memory allocator:                       system default
        Runtime Instrumentation (slow code):    No
        uuid support:                           Yes
        systemd support:                        Yes
        Config file:                            /etc/rsyslog.conf
        PID file:                               /var/run/rsyslogd.pid
        Number of Bits in RainerScript integers: 64

See https://www.rsyslog.com for more information.
[voxsteel@centos8 rsyslog.d]$ 
```

Figure 11.1 – Checking the rsyslog version

Check the status of the rsyslog service by running the `systemctl status rsyslog` command:

```
[voxsteel@centos8 rsyslog.d]$ systemctl status rsyslog
● rsyslog.service - System Logging Service
   Loaded: loaded (/usr/lib/systemd/system/rsyslog.service; enabled; vendor preset: enabled)
   Active: active (running) since Sun 2022-11-06 17:41:32 GMT; 6 days ago
     Docs: man:rsyslogd(8)
           https://www.rsyslog.com/doc/
 Main PID: 1564 (rsyslogd)
    Tasks: 4 (limit: 74384)
   Memory: 6.9M
   CGroup: /system.slice/rsyslog.service
           └─1564 /usr/sbin/rsyslogd -n
```

Figure 11.2 – Checking the status of the rsyslog service

As you can see, the service is active and running.

To check the status of the rsyslog service on multiple servers, you can use a configuration management tool such as Ansible or write a simple bash script to automate the process. Here's an example of how to check the rsyslog service on all servers using a bash script:

1. Create a file called `servers.txt` and add the list of servers you want to check, one per line:

    ```
    server1.example.com
    server2.example.com
    server3.example.com
    ```

2. Create a new bash script called `check_rsyslog_service.sh` and add the following code:

    ```
    #!/bin/bash
    while read server;
    do
    echo "Checking rsyslog service on $server"
    ssh $server "systemctl status rsyslog" ; done < servers.txt
    ```

3. Make the script executable:

    ```
    chmod +x check_rsyslog_service.sh
    ```

4. Run the script:

    ```
    ./check_rsyslog_service.sh
    ```

The script will iterate through the list of servers in `servers.txt` and execute the `systemctl status rsyslog` command over SSH to check the status of the rsyslog service on each server. The output will be displayed in the Terminal. You can modify the script to perform other actions on the servers, such as restarting the rsyslog service or updating the rsyslog configuration.

Configuring rsyslog for centralized logging

Centralized logging using the `central-rsyslog` server can be set up after you've updated to the most recent version of rsyslog.

The central logging setup is created by turning on the rsyslog UDP input module, `imudp`, and building the rsyslog template to receive log messages from other servers. The `imudp` input module allows syslog messages to be broadcast over UDP to be received by the `central-rsyslog` server.

Enable the options shown in the following screenshot in `/etc/rsyslog.conf` (the rsyslog configuration file) before saving the file and closing the editor.

The `imudp` input module needs to be configured to utilize the default UDP port of `514` to work:

```
#### MODULES ####

module(load="imuxsock"     # provides support for local system logging (e.g. via logger command)
       SysSock.Use="off") # Turn off message reception via local log socket;
                          # local messages are retrieved through imjournal now.
module(load="imjournal"            # provides access to the systemd journal
       StateFile="imjournal.state") # File to store the position in the journal
module(load="imklog") # reads kernel messages (the same are read from journald)
module(load="immark") # provides --MARK-- message capability

# Provides UDP syslog reception
# for parameters see http://www.rsyslog.com/doc/imudp.html
module(load="imudp") # needs to be done just once
input(type="imudp" port="514")

# Provides TCP syslog reception
# for parameters see http://www.rsyslog.com/doc/imtcp.html
#module(load="imtcp") # needs to be done just once
#input(type="imtcp" port="514")
```

Figure 11.3 – imudp module configuration

Then, create a new rsyslog template (`/etc/rsyslog.d/50-remote-logs.conf`) and paste the configuration indicated in *Figure 11.4*.

The following rsyslog template will allow you to collect logs from other servers and store them in the `/var/log/remotelogs/` directory on the `main-rsyslog` server:

```
[[voxsteel@centos8 rsyslog.d]$ cat 50-remote-logs.conf
# define template for remote loggin
# remote logs will be stored at /var/log/remotelogs directory
# each host will have specific directory based on the system %HOSTNAME%
# name of the log file is %PROGRAMNAME%.log such as sshd.log, su.log
# both %HOSTNAME% and %PROGRAMNAME% is the Rsyslog message properties
template (
    name="RemoteLogs"
    type="string"
    string="/var/log/remotelogs/%HOSTNAME%/%PROGRAMNAME%.log"
)

# gather all log messages from all facilities
# at all severity levels to the RemoteLogs template
*.* -?RemoteLogs

# stop the process once the file is written
stop
[voxsteel@centos8 rsyslog.d]$ █
```

Figure 11.4 – Template configuration

To establish a new log directory (/var/log/remotelogs/) owned by the root user with the adm group, run the following instructions. By doing this, the rsyslog service will be able to create logs in the /var/log/remotelogs folder:

```
mkdir -p /var/log/remotelogs
```

Then, change the ownership of the remotelogs folder:

```
chown -R root:adm /var/log/remotelogs/
```

To check the rsyslog settings (/etc/rsyslog.conf and /etc/rsyslog.d/50-remote-logs.conf), simply execute the rsyslogd commands provided here:

```
[voxsteel@centos8 rsyslog.d]$ rsyslogd -N1 -f /etc/rsyslog.d/50-remote-logs.conf
rsyslogd: version 8.2102.0-5.el8, config validation run (level 1), master config /etc/rsyslog.d/50-remote-logs.conf
rsyslogd: End of config validation run. Bye.
[voxsteel@centos8 rsyslog.d]$ █
```

Figure 11.5 – Checking the syntax

After double-checking the settings, you can restart the rsyslog service using the following command:

```
systemctl restart rsyslog
```

The rsyslog service, which has the imudp input module enabled, has exposed the syslog protocol's default UDP port, 514. Now, hosts can communicate with the main-rsyslog server by sending their logs there:

```
systemctl restart rsyslog
```

You can double-check that your ports have been properly opened by running the `ss` command, as follows:

```
ss -tulpn
```

Here's the output:

```
[voxsteel@centos8 rsyslog.d]$ ss -tulpn
Netid      State      Recv-Q      Send-Q              Local Address:Port          Peer Address:Port
udp        UNCONN     0           0                    0.0.0.0:5353                0.0.0.0:*
udp        UNCONN     0           0                    0.0.0.0:38553               0.0.0.0:*
udp        UNCONN     0           0              192.168.122.1:53                  0.0.0.0:*
udp        UNCONN     0           0             0.0.0.0%virbr0:67                  0.0.0.0:*
udp        UNCONN     0           0                    0.0.0.0:111                 0.0.0.0:*
udp        UNCONN     0           0                  127.0.0.1:323                 0.0.0.0:*
udp        UNCONN     0           0                  127.0.0.1:514                 0.0.0.0:*
udp        UNCONN     0           0                    0.0.0.0:43445               0.0.0.0:*
udp        UNCONN     0           0                       [::]:5353                   [::]:*
udp        UNCONN     0           0                       [::]:40552                  [::]:*
udp        UNCONN     0           0                       [::]:111                    [::]:*
udp        UNCONN     0           0                      [::1]:323                    [::]:*
tcp        LISTEN     0           128                  0.0.0.0:8085                0.0.0.0:*
tcp        LISTEN     0           32             192.168.122.1:53                  0.0.0.0:*
tcp        LISTEN     0           128                  0.0.0.0:8086                0.0.0.0:*
tcp        LISTEN     0           128                  0.0.0.0:22                  0.0.0.0:*
tcp        LISTEN     0           5                  127.0.0.1:631                 0.0.0.0:*
tcp        LISTEN     0           128                  0.0.0.0:9000                0.0.0.0:*
tcp        LISTEN     0           128                  0.0.0.0:3306                0.0.0.0:*
tcp        LISTEN     0           128                  0.0.0.0:111                 0.0.0.0:*
tcp        LISTEN     0           128                  0.0.0.0:8080                0.0.0.0:*
tcp        LISTEN     0           20                   0.0.0.0:80                  0.0.0.0:*
tcp        LISTEN     0           128                  0.0.0.0:8082                0.0.0.0:*
tcp        LISTEN     0           128                     [::]:8085                   [::]:*
tcp        LISTEN     0           128                     [::]:8086                   [::]:*
tcp        LISTEN     0           128                     [::]:22                     [::]:*
tcp        LISTEN     0           5                      [::1]:631                    [::]:*
tcp        LISTEN     0           128                        *:9090                      *:*
tcp        LISTEN     0           128                     [::]:9000                   [::]:*
tcp        LISTEN     0           128                     [::]:3306                   [::]:*
tcp        LISTEN     0           128                     [::]:111                    [::]:*
tcp        LISTEN     0           128                     [::]:8080                   [::]:*
tcp        LISTEN     0           128                     [::]:8082                   [::]:*
[voxsteel@centos8 rsyslog.d]$
```

Figure 11.6 – Command to see ports listening

Syslog is a simple and efficient protocol for collecting and managing log data in a distributed network environment. It provides a centralized location for storing logs, which makes it easier to manage, monitor, and troubleshoot systems.

Sending logs to a centralized rsyslog server

You've already taken the first step toward streamlined log handling by configuring syslog on the `main-rsyslog` server. But how do you know that the `main-rsyslog` server is receiving the logs? Logs can be sent from a remote client system to a `main-rsyslog` server by activating and configuring the rsyslog output module (`main-rsyslog`).

In this example, the `client01` machine uses the rsyslog output module, `omfwd`, to transmit logs to the `main-rsyslog` server.

To process messages and logs, the omfwd module must be installed (it will be already). It can be used in conjunction with rsyslog templates. Finally, the module uses the rsyslog action object to transmit the data through UDP and TCP to the specified destinations.

Set up the client machine so that it can submit logs to the main-rsyslog server.

Create a new rsyslog configuration (/etc/rsyslog.d/20-forward-logs.conf) in your preferred text editor and enter the settings shown in *Figure 11.7*.

Using the SendRemote template, log messages are formatted before being sent via the UDP protocol to the main-rsyslog server (192.168.1.111). In this case, the IP address should be replaced with the IP address of your primary rsyslog server:

```
[voxsteel@centos8 rsyslog.d]$ cat 20-forward-logs.conf
# process all log messages before sending
# with the SendRemote template
template(
    name="SendRemote"
    type="string"
    string="<%PRI%>%TIMESTAMP:::date-rfc3339% %HOSTNAME% %syslogtag:1:32%%msg:::sp-if-no-1st-sp%%msg%"
)

# forward log messages using omfwd module
# to the target server 172.16.1.10
# via UDP porotocol on port 514
# log messages is formatted using the SendRemote template
# setup queue for remote log
action(
    type="omfwd"
    Target="192.168.1.111"
    Port="514"
    Protocol="udp"
    template="SendRemote"

    queue.SpoolDirectory="/var/spool/rsyslog"
    queue.FileName="remote"
    queue.MaxDiskSpace="1g"
    queue.SaveOnShutdown="on"
    queue.Type="LinkedList"
    ResendLastMSGOnReconnect="on"
)

# stop process after the file is written
stop
[voxsteel@centos8 rsyslog.d]$
```

Figure 11.7 – Template for SendRemote

The preceding screenshot shows the content of a template file for log forwarding.

Check if the syntax is correct by running this command:

```
rsyslogd -N1 -f /etc/rsyslog.d/20-remote-logs.conf
```

Restart rsyslog by running the sudo systemctl restart rsyslog command and check whether the syslog server is receiving logs from the client.

Log rotation

Log rotation is a crucial process in Linux systems to manage log files efficiently. As applications and services generate log data over time, log files can grow significantly, consuming disk space and potentially leading to performance issues. Log rotation allows for the periodic compression, archival, and removal of old log files, ensuring the system maintains a manageable log history.

In Linux, log rotation is typically handled by a log rotation tool called `logrotate`. The configuration file for `logrotate` is located at `/etc/logrotate.conf`, and it includes references to individual log rotation configurations in the `/etc/logrotate.d/` directory.

Here's a step-by-step guide on how to configure log rotation in Linux:

1. **Install logrotate (if not already installed)**: Most Linux distributions come with `logrotate` pre-installed. However, if it's not available on your system, you can install it using the package manager specific to your Linux distribution. For example, on Debian/Ubuntu-based systems, you can install it with the following commands:

    ```
    sudo apt-get update
    sudo apt-get install logrotate
    ```

2. **Create a log rotation configuration file**: You can create a new log rotation configuration file for your specific application/service or use the default one. It's recommended to create separate files for different applications for easier management.

    Navigate to the `/etc/logrotate.d/` directory and create a new configuration file – for example, `myapp_logrotate`:

    ```
    sudo nano /etc/logrotate.d/myapp_logrotate
    ```

3. **Define the log rotation settings in the configuration file**: The `logrotate` configuration file follows a specific syntax. Here's a basic example:

    ```
    /path/to/your/logfile.log {
    rotate <N>       # Number of log files to keep before removal

    daily            # Frequency of rotation (daily, weekly, monthly,
    etc.)

    missingok        # Don't throw an error if the log file is missing

    notifempty       # Do not rotate an empty log file

    compress         # Compress the rotated log files using gzip

    create <mode> <user> <group> # Create new empty log file with
    ```

```
    specified permissions, user, and group


    }
```

Replace /path/to/your/logfile.log with the actual path to your log file. Replace <N> with the desired number of log files to keep before removal (for example, rotate 7 to keep 7 days' worth of logs). Replace <mode>, <user>, and <group> with the appropriate permissions and ownership for the newly created log file.

Save the configuration file and exit the text editor.

4. **Test the configuration**: To check if your logrotate configuration is error-free, you can run the following command:

```
sudo logrotate -d /etc/logrotate.d/myapp_logrotate
```

The -d flag is for debugging, and it will show you what logrotate would do without actually rotating the log files.

5. **Perform a manual log rotation**: Once you are confident that the configuration is correct, you can manually trigger log rotation with the following command:

```
sudo logrotate /etc/logrotate.d/myapp_logrotate
```

6. **Set up a cron job**: To automate log rotation, set up a cron job that runs logrotate at regular intervals. You can add an entry to crontab using the following command:

```
sudo crontab -e
```

Then, add the following line to run logrotate daily at midnight:

```
0 0 * * * /usr/sbin/logrotate /etc/logrotate.conf
```

Save crontab and exit the text editor.

Now, your log files will be automatically rotated and archived based on the configuration settings. You can adjust the rotation frequency and other options in the logrotate configuration file to suit your specific needs.

Journald

Journal is part of systemd. Messages from various parts of a systemd-enabled Linux machine are collected here. This comprises notifications from the kernel and boot process, syslog, and other services.

Traditionally, during Linux's boot process, the OS's many subsystems and application daemons would each log messages in text files. Different levels of detail would be logged for each subsystem's messages. When troubleshooting, administrators often had to sift through messages from several files spanning different periods and then correlate the contents. The journaling feature eliminates this problem by centrally logging all system and application-level messages.

The systemd-journald daemon is in charge of the journal. It gathers data from several resources and inserts the gathered messages into the diary.

When systemd is using in-memory journaling, the journal files are generated under the `/run/log/journal` folder. If there isn't already such a directory, one will be made. The journal is generated with persistent storage in the `/var/log/journal` directory; again, systemd will establish this directory if necessary. Logs will be written to `/run/log/journal` in a non-persistent fashion if this directory is destroyed; systemd-journald will not recreate it automatically. When the daemon is restarted, the directory is recreated.

The `journalctl` command is useful for debugging services and processes since it allows you to examine and modify the systemd logs.

The `journalctl` command and its numerous display options will be described next, along with how to view systemd logs. Since each machine has its own set of records, the results will vary.

To show all journal entries, use the `journalctl` command without any options:

```
[voxsteel@centos8 ~]$ journalctl
-- Logs begin at Wed 2023-01-18 14:07:03 GMT, end at Wed 2023-01-18
14:09:39 GMT. --
Jan 18 14:07:03 centos8 kernel: microcode: microcode updated early
to revision 0xd6, date = 2019-10-03Jan 18 14:07:03 centos8 kernel:
Linux version 4.18.0-348.7.1.el8_5.x86_64 (mockbuild@kbuilder.bsys.
centos.org) (gcc version 8.5.0 20210514 (Red Hat 8.5.0>Jan 18 14:07:03
centos8 kernel: Command line: BOOT_IMAGE=(hd1,gpt2)/vmlinuz-4.18.0-
348.7.1.el8_5.x86_64 root=/dev/mapper/cl-root ro crashkernel=auto
resu>Jan 18 14:07:03 centos8 kernel: x86/fpu: Supporting XSAVE
feature 0x001: 'x87 floating point registers'Jan 18 14:07:03 centos8
kernel: x86/fpu: Supporting XSAVE feature 0x002: 'SSE registers'Jan
18 14:07:03 centos8 kernel: x86/fpu: Supporting XSAVE feature 0x004:
'AVX registers'Jan 18 14:07:03 centos8 kernel: x86/fpu: Supporting
XSAVE feature 0x008: 'MPX bounds registers'Jan 18 14:07:03 centos8
kernel: x86/fpu: Supporting XSAVE feature 0x010: 'MPX CSR'Jan 18
14:07:03 centos8 kernel: x86/fpu: xstate_offset[2]:  576, xstate_
sizes[2]:  256Jan 18 14:07:03 centos8 kernel: x86/fpu: xstate_
offset[3]:  832, xstate_sizes[3]:   64Jan 18 14:07:03 centos8 kernel:
x86/fpu: xstate_offset[4]:  896, xstate_sizes[4]:   64Jan 18 14:07:03
centos8 kernel: x86/fpu: Enabled xstate_features 0x1f, context size
is 960 bytes, using 'compacted' format.Jan 18 14:07:03 centos8
kernel: BIOS-provided physical RAM map:Jan 18 14:07:03 centos8 kernel:
BIOS-e820: [mem 0x0000000000000000-0x0000000000057fff] usableJan
18 14:07:03 centos8 kernel: BIOS-e820: [mem 0x0000000000058000-
0x0000000000058fff] reserveJan 18 14:07:03 centos8 kernel: BIOS-
e820: [mem 0x0000000000059000-0x000000000009dfff] usableJan 18
14:07:03 centos8 kernel: BIOS-e820: [mem 0x000000000009e000-
0x000000000009efff] reservedJan 18 14:07:03 centos8 kernel: BIOS-
e820: [mem 0x000000000009f000-0x000000000009ffff] usableJan 18
14:07:03 centos8 kernel: BIOS-e820: [mem 0x00000000000a0000-
0x00000000000fffff] reservedJan 18 14:07:03 centos8 kernel: BIOS-
e820: [mem 0x0000000000100000-0x00000000c70fafff] usableJan 18
```

```
14:07:03 centos8 kernel: BIOS-e820: [mem 0x00000000c70fb000-
0x00000000c7c7efff] reservedJan 18 14:07:03 centos8 kernel: BIOS-e820:
[mem 0x00000000c7c7f000-0x00000000c7e7efff] ACPI NVS
```

The output shows the time range of the log data. The columns contain the following data in order from left to right:

- Date and time

- Host

- Log source

- Log message

To show logs specific to the current boot, use the -b tag, as follows:

```
[voxsteel@centos8 ~]$ journalctl -b
-- Logs begin at Wed 2023-01-18 14:07:03 GMT, end at Wed 2023-01-18
16:36:10 GMT. --
Jan 18 14:07:03 centos8 kernel: microcode: microcode updated early to
revision 0xd6, date = 2019-10-03
Jan 18 14:07:03 centos8 kernel: Linux version 4.18.0-348.7.1.el8_5.
x86_64 (mockbuild@kbuilder.bsys.centos.org) (gcc version 8.5.0
20210514 (Red Hat 8.5.0-4) (GCC)>
Jan 18 14:07:03 centos8 kernel: Command line: BOOT_IMAGE=(hd1,gpt2)/
vmlinuz-4.18.0-348.7.1.el8_5.x86_64 root=/dev/mapper/cl-root ro
crashkernel=auto resume=/dev/m>
Jan 18 14:07:03 centos8 kernel: x86/fpu: Supporting XSAVE feature
0x001: 'x87 floating point registers'
Jan 18 14:07:03 centos8 kernel: x86/fpu: Enabled xstate features 0x1f,
context size is 960 bytes, using 'compacted' format.
Jan 18 14:07:03 centos8 kernel: BIOS-provided physical RAM map:
Jan 18 14:07:03 centos8 kernel: BIOS-e820: [mem 0x0000000000000000-
0x0000000000057fff] usable
Jan 18 14:07:03 centos8 kernel: BIOS-e820: [mem 0x0000000000058000-
0x0000000000058fff] reserved
Jan 18 14:07:03 centos8 kernel: BIOS-e820: [mem 0x00000000c7eff000-
0x00000000c7efffff] usable
Jan 18 14:07:03 centos8 kernel: BIOS-e820: [mem 0x00000000c7f00000-
0x00000000cc7fffff] reserved
```

If you want to see the logs from the last 10 minutes, for example, then you can use journalctl -S "10 minutes ago":

```
[voxsteel@centos8 ~]$ journalctl -S "10 minutes ago"
-- Logs begin at Wed 2023-01-18 14:07:03 GMT, end at Wed 2023-01-18
16:38:00 GMT. --
Jan 18 16:31:49 centos8 systemd[1]: run-docker-runtime\x2drunc-moby-
586ec0d1511775a767ac92e0bc680e5ca772a18e59e31f9e358f9632834faede-runc.
```

```
ucvxT5.mount: Succeeded.
Jan 18 16:32:54 centos8 dbus-daemon[1048]: [system] Activating service
name='org.fedoraproject.Setroubleshootd' requested by ':1.30' (uid=0
pid=987 comm="/usr/sbi>
Jan 18 16:32:54 centos8 dbus-daemon[1048]: [system] Successfully
activated service 'org.fedoraproject.Setroubleshootd'
Jan 18 16:32:55 centos8 setroubleshoot[45450]: AnalyzeThread.run():
Cancel pending alarm
Jan 18 16:32:55 centos8 dbus-daemon[1048]: [system] Activating
service name='org.fedoraproject.SetroubleshootPrivileged' requested by
':1.1032' (uid=978 pid=45450>
Jan 18 16:32:56 centos8 dbus-daemon[1048]: [system] Successfully
activated service 'org.fedoraproject.SetroubleshootPrivileged'
Jan 18 16:32:57 centos8 setroubleshoot[45450]: SELinux is preventing /
usr/sbin/haproxy from name_connect access on the tcp_socket port 8082.
For complete SELinux >
Jan 18 16:32:57 centos8 setroubleshoot[45450]: SELinux is preventing /
usr/sbin/haproxy from name_connect access on the tcp_socket port 8082.
```

If you want to display only kernel journal log messages, then use the -k option, as follows:

```
[voxsteel@centos8 ~]$ journalctl -k
-- Logs begin at Wed 2023-01-18 14:07:03 GMT, end at Wed 2023-01-18
16:46:01 GMT. --
Jan 18 14:07:03 centos8 kernel: microcode: microcode updated early to
revision 0xd6, date = 2019-10-03
Jan 18 14:07:03 centos8 kernel: Linux version 4.18.0-348.7.1.el8_5.
x86_64 (mockbuild@kbuilder.bsys.centos.org) (gcc version 8.5.0
20210514 (Red Hat 8.5.0-4) (GCC)>
Jan 18 14:07:03 centos8 kernel: Command line: BOOT_IMAGE=(hd1,gpt2)/
vmlinuz-4.18.0-348.7.1.el8_5.x86_64 root=/dev/mapper/cl-root ro
crashkernel=auto resume=/dev/m>
Jan 18 14:07:03 centos8 kernel: x86/fpu: Supporting XSAVE feature
0x001: 'x87 floating point registers'
Jan 18 14:07:03 centos8 kernel: x86/fpu: Supporting XSAVE feature
0x002: 'SSE registers'
Jan 18 14:07:03 centos8 kernel: x86/fpu: Supporting XSAVE feature
0x004: 'AVX registers'
```

You can also filter log messages based on priority using the following command:

```
journalctl -p <number or text priority>
```

The following are the priorities levels:

- **Emergency**: 0 or emerg
- **Alert**: 1 or alert
- **Critical**: 2 or crit

- **Error**: 3 or `err`

- **Warning**: 4 or `warning`

- **Notice**: 5 or `notice`

- **Inform**: 6 or `info`

- **Debug**: 7 or `debug`

You can find all the parameters available for `journalctl` using the `man journalctl` command.

DMESG

dmesg is a command-line tool in Linux that allows you to view the kernel ring buffer messages. The kernel ring buffer is a circular buffer in memory that stores messages generated by the kernel, such as hardware events, device driver information, and system error messages.

The `dmesg` command displays the contents of this kernel ring buffer, allowing you to view messages that have been generated since the system was last booted. These messages can be useful for debugging system problems, identifying hardware issues, and monitoring system events.

Some of the common use cases of `dmesg` include the following:

- **Troubleshooting system issues**: `dmesg` can be used to identify and diagnose system problems by displaying error messages, warnings, and other relevant information

- **Checking hardware status**: `dmesg` can provide information about hardware devices and drivers, such as when a device is detected or when a driver fails to load

- **Monitoring system events**: `dmesg` can be used to monitor system events, such as when a user plugs in a USB device or when a system service starts or stops

Here are some commonly used options that are used with the `dmesg` command:

- `-T`: Displays the timestamp in human-readable format

- `-H`: Displays the output in a more human-readable format

- `-l level`: Displays only messages of the specified log level (`debug`, `info`, `notice`, `warning`, `err`, `crit`, `alert`, or `emerg`)

- `-k`: Displays only kernel messages

Overall, `dmesg` is a powerful tool that can help you troubleshoot system problems and monitor system events in Linux.

The `dmesg` command provides a window into the inner workings of Linux. This *fault finder's friend* allows you to read and observe messages sent by the kernel's hardware devices and drivers from the kernel's internal ring buffer.

Understanding the ring buffer in Linux

When a computer is powered on, several events occur in a specific order; in Linux and Unix-like systems, these activities are referred to as booting and startup, respectively.

After the initialization of the system has been completed by the boot procedures (BIOS or UEFI, MBR, and GRUB), the kernel is loaded into memory, the initial ramdisk (initrd or initramfs) is connected to the kernel, and systemd is launched.

The OS is handed over to the startup routines, which finish the setup. When a system is first booted, it may take a while for logging daemons such as syslogd and rsyslogd to become operational. The kernel features a ring buffer that it employs as a message cache to ensure that critical error messages and warnings from this phase of initialization are not lost.

A ring buffer is a special area of memory where messages can be stored. It has a standard size and straightforward construction. When it reaches capacity, newer messages replace older ones. It can be seen conceptually as a *circular buffer*.

Information such as device driver initialization messages, hardware messages, and kernel module messages are all kept in the kernel ring buffer. The ring buffer is a handy place to begin troubleshooting hardware faults or other startup issues because it stores these low-level messages.

With the dmesg command, you can examine the log of messages saved in the system's ring buffer:

```
dmesg -T | less
```

I added -T to show the timestamps in a readable format, and less to make it scrollable.

The following is the output of the preceding command:

```
[Fri Jan 20 08:12:36 2023] wlp2s0: associate with d4:5d:64:e1:e0:2c
(try 1/3)
[Fri Jan 20 08:12:36 2023] wlp2s0: RX AssocResp from d4:5d:64:e1:e0:2c
(capab=0x1011 status=0 aid=5)
[Fri Jan 20 08:12:36 2023] wlp2s0: associated
[Fri Jan 20 08:12:36 2023] wlp2s0: Limiting TX power to 23 (23 - 0)
dBm as advertised by d4:5d:64:e1:e0:2c
[Fri Jan 20 15:08:39 2023] atkbd serio0: Unknown key pressed
(translated set 2, code 0x85 on isa0060/serio0).
[Fri Jan 20 15:08:39 2023] atkbd serio0: Use 'setkeycodes e005
<keycode>' to make it known.
[Mon Jan 23 06:27:58 2023] wlp2s0: deauthenticating from
d4:5d:64:e1:e0:2c by local choice (Reason: 2=PREV_AUTH_NOT_VALID)
[Mon Jan 23 06:27:59 2023] wlp2s0: authenticate with d4:5d:64:e1:e0:2c
[Mon Jan 23 06:27:59 2023] wlp2s0: send auth to d4:5d:64:e1:e0:2c (try
1/3)
[Mon Jan 23 06:27:59 2023] wlp2s0: authenticated
[Mon Jan 23 06:27:59 2023] wlp2s0: associate with d4:5d:64:e1:e0:2c
```

```
(try 1/3)
[Mon Jan 23 06:27:59 2023] wlp2s0: RX AssocResp from d4:5d:64:e1:e0:2c
(capab=0x1011 status=0 aid=5)
[Mon Jan 23 06:27:59 2023] wlp2s0: associated
[Mon Jan 23 06:27:59 2023] wlp2s0: Limiting TX power to 23 (23 - 0)
dBm as advertised by d4:5d:64:e1:e0:2cresume=/dev/mapper/cl-swap
rd.lvm.lv=cl/root rd.lvm.lv=cl/swap rhgb quiet
[Wed Jan 18 14:06:39 2023] x86/fpu: Supporting XSAVE feature 0x001:
'x87 floating point registers'
```

The dmesg --follow command is a variation of the dmesg command that continuously displays new messages as they are generated in the kernel ring buffer.

When you run dmesg --follow in a Terminal, it will display the most recent kernel messages and then continue to display any new messages that are generated in real time. This can be useful for monitoring system events as they occur or for diagnosing issues that may be occurring in real time.

The --follow option is equivalent to the -w or --wait option, which tells dmesg to wait for new messages to be added to the kernel ring buffer and display them as they come in.

Here are some use cases for the dmesg --follow command:

- **Monitoring hardware events**: If you're troubleshooting a hardware issue, you can use dmesg --follow to monitor the kernel messages as you plug or unplug devices, or as you interact with hardware

- **Debugging system issues**: If you're trying to diagnose a system issue, dmesg --follow can help you see what's happening in real time and identify any patterns or issues that may be causing the problem

- **Monitoring system health**: If you want to keep an eye on your system's health, you can use dmesg --follow to watch for any error messages or warnings that may be generated in the kernel ring buffer

It's worth noting that because dmesg --follow continuously displays new messages, the output can quickly become overwhelming and difficult to read. To stop the dmesg command from running, you can press *Ctrl + C* in the terminal:

```
dmesg -follow
[151557.551942] wlp2s0: Limiting TX power to 23 (23 - 0) dBm as
advertised by d4:5d:64:e1:e0:2c
[176520.971449] atkbd serio0: Unknown key pressed (translated set 2,
code 0x85 on isa0060/serio0).
[176520.971452] atkbd serio0: Use 'setkeycodes e005 <keycode>' to make
it known.
[404479.355923] wlp2s0: deauthenticating from d4:5d:64:e1:e0:2c by
local choice (Reason: 2=PREV_AUTH_NOT_VALID)
[404480.713565] wlp2s0: authenticate with d4:5d:64:e1:e0:2c
```

```
[404480.722235] wlp2s0: send auth to d4:5d:64:e1:e0:2c (try 1/3)
[404480.724148] wlp2s0: authenticated
[404480.724865] wlp2s0: associate with d4:5d:64:e1:e0:2c (try 1/3)
[404480.725868] wlp2s0: RX AssocResp from d4:5d:64:e1:e0:2c
(capab=0x1011 status=0 aid=5)
[404480.727602] wlp2s0: associated
[404480.781339] wlp2s0: Limiting TX power to 23 (23 - 0) dBm as
advertised by d4:5d:64:e1:e0:2c
```

Observe that you are not taken back to the prompt where you entered commands. Whenever fresh messages are received, dmesg will show them in the Terminal's footer.

You can see the last 15 messages, for example, by using the following command:

```
dmesg | tail -15
```

Alternatively, you can search for specific terms (for example, memory) using the dmesg | grep -I memory command:

```
[    0.000000] Memory: 2839388K/12460644K available (12293K kernel
code, 2261K rwdata, 7872K rodata, 2492K init, 13944K bss, 795924K
reserved, 0K cma-reserved)
[    0.021777] Freeing SMP alternatives memory: 32K
[    0.048199] x86/mm: Memory block size: 128MB
[    4.052723] Freeing initrd memory: 53084K
[    4.215935] Non-volatile memory driver v1.3
[    6.181994] Freeing unused decrypted memory: 2036K
[    6.182301] Freeing unused kernel memory: 2492K
[    6.188636] Freeing unused kernel memory: 2012K
[    6.188833] Freeing unused kernel memory: 320K
[    8.302610] i915 0000:00:02.0: [drm] Reducing the compressed
framebuffer size. This may lead to less power savings than a
non-reduced-size. Try to increase stolen memory size if available in
BIOS.
[   37.829370] PM: Saving platform NVS memory
[   37.837899] PM: Restoring platform NVS memory
```

Use the man command to find out all the magic that you can do with one command. dmesg is a very powerful tool for investigating logs.

Summary

In this chapter, you learned about logs and how to configure rsyslog for centralized logging. Using various rsyslog input and output plugins, you transmitted server logs over the network to the consolidated rsyslog server. Your rsyslog server is now the only location you need to look for logs.

We also provided examples of how to read systemd journal logs. The `journalctl` command is a powerful resource for diagnosing issues with Linux services and finding problems in the OS.

Finally, you learned about the power of the `dmesg` command and how it can be used. `dmesg` is a powerful tool that can help you troubleshoot system problems and monitor system events in Linux.

In the next chapter, we will talk about centralized authentication, where you can use a single server for all your clients to authenticate against.

12
Centralized Authentication

User access control is a critically important part of information security. On a single machine, keeping track of users and making sure only authorized people have access is simple, but as networks become larger, it becomes increasingly difficult to keep user accounts in sync on all machines, which is why large networks use centralized authentication mechanisms. Historically, UNIX-like systems usually used **Network Information Service (NIS)**, developed by Sun Microsystems – a once widespread but now mostly unused protocol. These days, the choice is wider and includes standalone LDAP directories, Kerberos realms, or authentication solutions that provide a combination of a directory service for storing user information and single sign-on protocols, such as FreeIPA and Microsoft Active Directory.

In this chapter, we will learn about the following:

- Authentication and user information lookup frameworks in Linux
- The roles of the **Name Service Switch (NSS)** framework, **Pluggable Authentication Modules (PAM)**, and the **System Security Services Daemon (SSSD)**
- How to set up a domain controller that's compatible with Microsoft Active Directory and connect a client machine to it

The AAA framework

The access control framework is often referred to as *AAA* due to its three components: *authentication*, *authorization*, and *accounting*.

Authentication is responsible for verifying the user's identity – usually by checking whether the user possesses certain knowledge (such as a password), data (such as a cryptographic key or the correct seed for a time-based authentication algorithm), a physical item (such as a hardware key storage), or an attribute (such as a fingerprint).

Authorization is the process of checking whether the user that attempts to execute an action has permission to do so. Since in UNIX systems many entities, such as hardware devices and sockets, are represented as files, a lot of the time, file access permissions are used as an authorization framework.

Finally, the **accounting** process ensures that user actions are recorded so that it is possible to attribute actions to users, monitor user activity for anomalies, and investigate security incidents. Since, for a general-purpose OS, it is impossible to give an exhaustive list of user actions, there cannot be a general accounting framework. The syslog mechanism is a common way to record log messages, but the log message format is different for each application.

Among the user access control components, authentication in Linux is unique in that there is a widely used and general framework for it that consists of three parts: **NSS**, **PAM**, and, on newer Linux distributions, **SSSD**. The relationship between them is somewhat complicated because their functionality is broad and partially overlaps, and many tasks can be solved at different levels.

The two older parts of that framework, NSS and PAM, originated in the OS named Solaris, which was developed by Sun Microsystems (later acquired by Oracle), and were soon adopted by almost all UNIX-like systems. However, neither mechanism became a part of the POSIX standard and its implementations in different OSs are slightly different. SSSD was developed for Linux in the late 2000s and is now widely used by Linux distributions, but not by other UNIX-like systems.

Let's examine the purpose and functionality of those subsystems in detail.

Authentication mechanisms in Linux

Before we learn about centralized authentication mechanisms, we need to learn how authentication works in Linux in general. Before a system can check user credentials, it needs to fetch user information first – let's examine how information lookup works.

Information lookup

Information about users and groups is necessary for authentication, but it has many other uses. For example, file ownership information is usually displayed in a human-readable fashion, but internally, filesystems store numeric user and group identifiers, so even programs that have nothing to do with security, such as `ls`, may need to have a way to look up information about users and groups to map their identifiers to names.

The POSIX API standard includes multiple functions for querying information about various entities such as users, groups, and hosts. For example, `gethostbyname(name)` retrieves network addresses associated with a domain name, and `getgrpnam(name)` retrieves information about a user group. For applications, those functions are black boxes, and they have neither knowledge nor control of the process – where and how the underlying system gets that data is not an application developer's concern. By keeping the process opaque to applications, OS developers can make sure that all applications get consistent data from lookup calls.

In GNU/Linux systems, those functions are implemented by the GNU C library (`glibc`). That library, however, does not have a fixed set of supported lookup mechanisms and data sources – instead, it uses the **NSS** mechanism as an abstraction layer.

Name Service Switch

The NSS subsystem allows an administrator to map *databases* with different types of information to *sources*. Let's inspect its configuration file from a Fedora installation to get a sense of what it is capable of:

```
$ cat /etc/nsswitch.conf


# Generated by authselect on Tue May 31 00:21:30 2022
# Do not modify this file manually, use authselect instead. Any user
changes will be overwritten.
# You can stop authselect from managing your configuration by calling
'authselect opt-out'.
# See authselect(8) for more details.
# In order of likelihood of use to accelerate lookup.
passwd:        files sss systemd
shadow:        files
group:         files sss systemd
hosts:         files myhostname mdns4_minimal [NOTFOUND=return] resolve
[!UNAVAIL=return] dns
services:      files sss
netgroup:      files sss
automount:     files sss
aliases:       files
ethers:        files
gshadow:       files
networks:      files dns
protocols:     files
publickey:     files
rpc:           files
```

NSS is not limited to authentication data lookup and its functionality is much broader. The concept of NSS databases and sources is easier to demonstrate on a non-authentication example: hostname lookup. The `hosts` database, which is responsible for network hostname resolution, is set to use multiple sources, including `files` and `dns`.

The order of sources determines the order of lookup. In this case, the entry for `hosts` tells the system to look up names in local files first, and then turn to network sources such as zero-configuration network discovery mechanisms (multicast DNS), and finally to the global DNS system. In this particular case, the `files` keyword refers to the `/etc/hosts` file, which was discussed in *Chapter 6, Basic System Settings*.

Sources generally come in order from the fastest and most reliable. Local sources are placed first because they are always available even when the network is down and reading a local file is a very fast operation. Remote sources come later because accessing them always involves at least some latency and they may become inaccessible due to network or server faults. Sometimes, like in the case of hostname lookup, the order also has security implications: making the system look up names in /etc/hosts before making a request to a DNS server ensures that traditional names such as localhost will always point to the localhost (127.0.0.1 for IPv4 or ::1 for IPv6). If DNS were consulted first, a malicious or misconfigured server could return a different address and redirect traffic from commands such as ping localhost to an arbitrary address.

The passwd database is used to look up user information. Its files source is the familiar /etc/passwd – that is where its name comes from.

Lookup in different sources is implemented in dynamically loaded libraries (shared objects) such as /usr/lib/libnss_files.so.2 and /usr/lib/libnss_dns.so.2.

It's certainly possible to implement support for a different authentication information database using NSS alone and reuse the oldest login-handling code for it – the goal of NSS is to make such client code work with any source so that the code will not even know that it now gets user information and password hashes from a remote source such as a RADIUS or LDAP server rather than from /etc/passwd and /etc/shadow. That was done in the past and NSS modules for LDAP can still be found on the internet.

However, there are multiple reasons for a more flexible user identity-checking framework and an abstraction layer that provides authentication and lookup functionality.

Pluggable Authentication Modules

NSS helps programs retrieve various information, including usernames, group membership information, and password hashes. However, the logic for authentication still has to exist somewhere. For example, to conduct password-based authentication, there must be code that calculates a hash sum from a plain text password that the user enters and checks it against the hash stored in an authentication database.

However, there is more to authentication policies than just having passwords and checking that they are correct. Administrators may want to enforce password-strength rules or use multi-factor authentication to increase security, for example. Authentication using remote databases also presents challenges, such as credential caching to ensure that users are not locked out of their machines when the remote database becomes temporarily unavailable.

To allow developers and administrators to create and set up tools for flexible authentication policies and add new authentication algorithms, Linux distributions use a framework named **PAM**.

PAM provides an API for applications to authenticate users and for security tool developers to implement authentication mechanisms. PAM modules can either rely on the NSS layer to look up the information they need to authenticate a user or provide their own lookup mechanisms and configuration files.

On Fedora and Red Hat systems, PAM modules can be found in `/usr/lib64/security/`, while on Debian (for x86 machines), they can be found in `/usr/lib/x86_64-linux-gnu/security/`. Modules typically come with their own manual pages, so it is possible to get a brief description by running `man pam_unix` or `man pam_empty`.

For example, `pam_empty` is the simplest possible module that always returns an authentication failure – it is only useful as a sentinel module to ensure that when none of the multiple fallback mechanisms succeed, the user is denied access.

The `pam_unix` module implements the usual password authentication that normally uses the `/etc/passwd` and `/etc/shadow` files unless NSS is configured to use anything else.

Some modules do not implement any authentication mechanisms but rather perform auxiliary actions. For example, `pam_motd` shows a message (commonly found in `/etc/motd`) after login; `pam_mail` is what's responsible for checking for local email messages and displaying **You have new mail** if there are any, and modules such as `pam_pwquality` and `pam_pwhistory` help ensure that users do not use weak passwords and do not reuse their old passwords.

PAM configuration

While configuring PAM by hand is rarely a good idea and most Linux distributions discourage it and provide high-level configuration tools instead, it is still important to understand how its configuration files work.

First of all, PAM is not a program, but a framework and a set of APIs that are used by other programs. There is no single login program in Linux, so there is no one global authentication configuration file either. Instead, there are multiple programs with login capabilities that share most but not all of the configurations. These programs include the `/bin/login` executable, which handles local virtual console login attempts, but PAM is also independently used by graphical login managers (such as GDM or LightDM), screensavers, and network access services (such as OpenSSH).

Configuration files are stored in `/etc/pam.d/`, but none of them have any special meaning by themselves – they are all read and used by different programs. For example, the file named `/etc/pam.d/login` is used by `/bin/login` and thus applied only to local virtual consoles.

The names of those application-specific files are hardcoded in the programs that use them, but the shared configuration is stored in separate files whose names vary from one distribution to another.

Let's compare default configuration files for the OpenSSH daemon's use of PAM on Fedora and Debian. If you don't have OpenSSH installed, you can check out a different file, such as `/etc/pam.d/login`.

This file is from Fedora:

```
$ cat /etc/pam.d/sshd
#%PAM-1.0
auth       substack     password-auth
auth       include      postlogin
account    required     pam_sepermit.so
account    required     pam_nologin.so
account    include      password-auth
password   include      password-auth
# pam_selinux.so close should be the first session rule
session    required     pam_selinux.so close
session    required     pam_loginuid.so
# pam_selinux.so open should only be followed by sessions to be
executed in the user context
session    required     pam_selinux.so open env_params
session    required     pam_namespace.so
session    optional     pam_keyinit.so force revoke
session    optional     pam_motd.so
session    include      password-auth
session    include      postlogin
```

And this one is from a Debian system:

```
$ cat /etc/pam.d/sshd
# PAM configuration for the Secure Shell service
# Standard Un*x authentication.
@include common-auth
# Disallow non-root logins when /etc/nologin exists.
account    required     pam_nologin.so
# Uncomment and edit /etc/security/access.conf if you need to set
complex
# access limits that are hard to express in sshd_config.
# account  required     pam_access.so
# Standard Un*x authorization.
@include common-account
# SELinux needs to be the first session rule.  This ensures that any
# lingering context has been cleared.  Without this it is possible
# that a module could execute code in the wrong domain.
session [success=ok ignore=ignore module_unknown=ignore
default=bad]          pam_selinux.so close
# Set the loginuid process attribute.
session    required     pam_loginuid.so
# Create a new session keyring.
session    optional     pam_keyinit.so force revoke
```

```
# Standard Un*x session setup and teardown.
@include common-session
# Print the message of the day upon successful login.
# This includes a dynamically generated part from /run/motd.dynamic
# and a static (admin-editable) part from /etc/motd.
session    optional    pam_motd.so  motd=/run/motd.dynamic
session    optional    pam_motd.so noupdate
# Print the status of the user's mailbox upon successful login.
session    optional    pam_mail.so standard noenv # [1]
# Set up user limits from /etc/security/limits.conf.
session    required    pam_limits.so
# Read environment variables from /etc/environment and
# /etc/security/pam_env.conf.
session    required    pam_env.so # [1]
# In Debian 4.0 (etch), locale-related environment variables were
# moved to
# /etc/default/locale, so read that as well.
session    required    pam_env.so user_readenv=1 envfile=/etc/
default/locale
# SELinux needs to intervene at login time to ensure that the process
# starts
# in the proper default security context.  Only sessions which are
# intended
# to run in the user's context should be run after this.
session [success=ok ignore=ignore module_unknown=ignore
default=bad]          pam_selinux.so open
# Standard Un*x password updating.
@include common-password
```

As you can see, they are nearly identical in their functionality but organized differently. For example, Debian makes extensive use of the @include directive, which loads all lines from a different file. For example, @include common-auth instructs PAM to load /etc/pam.d/common-auth – all such file references are relative to /etc/pam.d.

The authors of the Fedora file chose to use a module interface call instead: auth substack system-auth. In such lines, the last option is either a module name or a configuration file reference. If it ends with .so (the usual extension of shared objects – dynamically loadable library files), then it is a module; otherwise, it's a configuration file in /etc/pam.d. So, when you see substack system-auth, it means loading lines from /etc/pam.d/system-auth (the include and substack options work similarly).

The first part of all such lines is a module interface name. Some modules provide only one interface type (for example, pam_unix only handles password authentication), while others may provide more than one type. The auth interface is for performing authentication. Other interface types show that *Pluggable Authentication Modules'* name is partially a misnomer. The account interface handles authorization and checks whether the user's account is allowed to log in – for example, the `pam_time` module can allow specific users to access the system only at certain hours or on certain days. The `password` interface handles password changing. Finally, the `session` interface handles auxiliary tasks unrelated to authentication, such as creating home directories for users on their first login or mounting directories over the network.

Limitations of PAM

PAM is only concerned with authentication and related activities; so, for example, configuring a `pam_ldap` module with the correct LDAP server and options will not automatically make user and group information from LDAP available to all applications that need to map numeric identifiers to names. In that case, to provide a seamless experience to users, an administrator also needs an NSS module. Configuring two modules with the same information independently creates a significant maintenance burden, and feeding information from remote sources to a simple module such as `pam_unix` does not work for single sign-on protocols such as Kerberos that require issuance and verification of session tokens rather than simple password hash sum checking. Historically, that problem was solved with custom daemons such as `winbind` for interaction with Microsoft Active Directory or Samba domains. These days, most Linux distributions solve this with a general abstraction layer – the **SSSD**.

System Security Services Daemon

The combination of NSS and PAM allows great flexibility but can also make common scenarios hard to configure and maintain. The SSSD project strives to simplify that process by coordinating the interaction of both PAM and NSS with remote databases.

One source of configuration complexity for single sign-on schemes is that they usually involve multiple components and protocols, such as LDAP for storing user information and Kerberos for issuing and checking cryptographic authentication tickets, plus a way to discover those services, typically via special DNS records. SSSD has built-in support for popular SSO schemes such as Microsoft Active Directory and FreeIPA, which greatly simplifies the process.

For this demonstration, we will set up a Microsoft Active Directory-compatible domain controller on Linux using the Samba project and then make a client machine join its domain. We will use Fedora Linux on both, but other distributions would mostly differ just in package installation commands.

Active Directory authentication with Samba 4

Samba is an open source implementation of multiple protocols that are required for interoperability with Microsoft Windows machines, including the SMB file-sharing protocol (which is where the Samba name came from). Apart from file sharing, it also implements authentication and user management – initially, it only supported Windows NT domains, but since 4.0.0, it has full support for Active Directory that's compatible with Windows Server 2008 and also includes built-in LDAP and DNS backends, which makes small installations very simple to deploy.

Setting up the domain controller

First, you will need to install the Samba domain controller package:

```
$ sudo dnf install samba-dc
```

Then, you may want to remove all configuration files for Samba and the Kerberos daemon to ensure a clean state:

```
$ sudo rm -f /etc/samba/smb.conf
$ sudo rm -f /etc/krb5.conf
```

Samba includes a command for automatically provisioning domain controllers, so there is no need to write configuration files by hand. This command supports both interactive and non-interactive modes, and it is also possible to specify some parameters through command-line options but enter the rest interactively. For example, if you plan to run the client system in a VirtualBox VM, you can set the controller to only listen on the VM network interface with `--option="interfaces=vboxnet0"` `--option="bind interfaces only=yes"`. You may also want to include fields for UNIX users in the LDAP schema that the controller will use – they are defined by RFC2307, so that option is `--use-rfc2307`.

The most important required parameter is the realm name, which must be a domain name written in capital letters – this is a requirement of the Kerberos protocol, and lowercase realm names will be rejected. We will use `AD.EXAMPLE.COM`, where `example.com` is one of the domains reserved for examples and documentation that is guaranteed not to belong to any real person or organization. Most modern client software uses the realm name primarily or exclusively, but the provision utility will also ask for the domain name (in the NetBIOS rather than DNS sense) – a string up to 15 characters long. Since Active Directory and its compatible implementations rely on DNS SRV records to find controllers and communicate with them, the controller will also serve those DNS records, but we will limit it to listen on the VM network interface – in this case, `192.168.56.1` (the default address for host-only network adapters in VirtualBox). It will also require a domain administrator password – you need to remember it to join that domain from client machines:

```
$ sudo samba-tool domain provision --option="interfaces=vboxnet0"
--option="bind interfaces only=yes" --use-rfc2307 --interactive
Realm:  AD.EXAMPLE.COM
Domain [AD]:  AD
Server Role (dc, member, standalone) [dc]:
DNS backend (SAMBA_INTERNAL, BIND9_FLATFILE, BIND9_DLZ, NONE) [SAMBA_
INTERNAL]:
DNS forwarder IP address (write 'none' to disable forwarding)
[127.0.0.53]:  192.168.56.1
Administrator password:
Retype password:
INFO 2022-12-07 16:25:13,958 pid:54185 /usr/lib64/python3.10/site-
packages/samba/provision/__init__.py #2108: Looking up IPv4 addresses
```

```
INFO 2022-12-07 16:25:17,450 pid:54185 /usr/lib64/python3.10/site-
packages/samba/provision/__init__.py #2017: Fixing provision GUIDs
...
INFO 2022-12-07 16:25:17,784 pid:54185 /usr/lib64/python3.10/
site-packages/samba/provision/__init__.py #2342: The Kerberos KDC
configuration for Samba AD is located at /var/lib/samba/private/kdc.
conf
INFO 2022-12-07 16:25:17,785 pid:54185 /usr/lib64/python3.10/site-
packages/samba/provision/__init__.py #2348: A Kerberos configuration
suitable for Samba AD has been generated at /var/lib/samba/private/
krb5.conf
INFO 2022-12-07 16:25:17,785 pid:54185 /usr/lib64/python3.10/site-
packages/samba/provision/__init__.py #2350: Merge the contents of this
file with your system krb5.conf or replace it with this one. Do not
create a symlink!
INFO 2022-12-07 16:25:17,948 pid:54185 /usr/lib64/python3.10/site-
packages/samba/provision/__init__.py #2082: Setting up fake yp server
settings
INFO 2022-12-07 16:25:18,002 pid:54185 /usr/lib64/python3.10/site-
packages/samba/provision/__init__.py #487: Once the above files are
installed, your Samba AD server will be ready to use
INFO 2022-12-07 16:25:18,002 pid:54185 /usr/lib64/python3.10/
site-packages/samba/provision/__init__.py #492: Server
Role:          active directory domain controller
INFO 2022-12-07 16:25:18,002 pid:54185 /usr/lib64/
python3.10/site-packages/samba/provision/__init__.py #493:
Hostname:           mizar
INFO 2022-12-07 16:25:18,002 pid:54185 /usr/lib64/python3.10/site-
packages/samba/provision/__init__.py #494: NetBIOS Domain:        AD
INFO 2022-12-07 16:25:18,002 pid:54185 /usr/lib64/python3.10/site-
packages/samba/provision/__init__.py #495: DNS Domain:           ad.
example.com
INFO 2022-12-07 16:25:18,002 pid:54185 /usr/lib64/python3.10/
site-packages/samba/provision/__init__.py #496: DOMAIN
SID:            S-1-5-21-2070738055-845390856-828010526
```

Note that the provisioning script generates a configuration file for Kerberos but does not deploy it – you need to copy it to its target location yourself:

```
$ sudo cp /var/lib/samba/private/krb5.conf /etc/krb5.conf
```

Now, all we need to do is start the Samba service:

```
$ sudo systemctl start samba
```

We will also create a user named `testuser` to verify that authentication is working as expected. The command to create a new user in our controller is `sudo samba-tool user create testuser` (it will ask you to enter a password).

With `sudo samba-tool user show testuser`, you can see the LDAP fields for that user account:

```
$ samba-tool user show testuser
dn: CN=testuser,CN=Users,DC=ad,DC=example,DC=com
objectClass: top
objectClass: person
objectClass: organizationalPerson
objectClass: user
cn: testuser
instanceType: 4
whenCreated: 20221207171435.0Z
uSNCreated: 4103
name: testuser
objectGUID: 773d2799-e08c-41fb-9768-675dd59edbbb
badPwdCount: 0
codePage: 0
countryCode: 0
badPasswordTime: 0
lastLogoff: 0
primaryGroupID: 513
objectSid: S-1-5-21-2070738055-845390856-828010526-1103
accountExpires: 9223372036854775807
sAMAccountName: testuser
sAMAccountType: 805306368
userPrincipalName: testuser@ad.example.com
objectCategory:
CN=Person,CN=Schema,CN=Configuration,DC=ad,DC=example,DC=com

pwdLastSet: 133149068751598230
userAccountControl: 512
lastLogonTimestamp: 133149085286105320
lockoutTime: 0
whenChanged: 20221207180044.0Z
uSNChanged: 4111
logonCount: 0
distinguishedName: CN=testuser,CN=Users,DC=ad,DC=example,DC=com
```

The server is now ready to receive authentication requests.

Setting up the client

Now that the domain controller is ready, we can set up a client machine to authenticate users against it. Since we are using Fedora for the client as well, it most likely has both SSSD and `realmd` – a tool for configuring authentication realms – installed by default.

We can verify that the controller is working using a `realm discover` command, as follows:

```
$ realm discover ad.example.com
ad.example.com
  type: kerberos
  realm-name: AD.EXAMPLE.COM
  domain-name: ad.example.com
  configured: no
  server-software: active-directory
  client-software: sssd
  required-package: oddjob
  required-package: oddjob-mkhomedir
  required-package: sssd
  required-package: adcli
  required-package: samba-common-tools
```

From this output, we can see that this machine has not joined that domain yet (`configured: no`). It also shows a list of packages that need to be installed to join it – if any are not installed yet, you can install them with a DNF command such as `sudo dnf install samba-common-tools`.

Joining the domain is also very simple. Once you run `sudo realm join ad.example.com`, it will ask you for the domain administrator password (the one you entered on the controller during the provisioning process) and generate configuration files for SSSD:

```
$ sudo realm join ad.example.com
Password for Administrator:
```

You will need to restart SSSD with `sudo systemctl restart sssd` to apply the changes.

The generated SSSD configuration will look like this:

```
$ sudo cat /etc/sssd/sssd.conf
[sssd]
domains = ad.example.com
config_file_version = 2
services = nss, pam
[domain/ad.example.com]
default_shell = /bin/bash
krb5_store_password_if_offline = True
cache_credentials = True
krb5_realm = AD.EXAMPLE.COM
```

```
realmd_tags = manages-system joined-with-adcli
id_provider = ad
fallback_homedir = /home/%u@%d
ad_domain = ad.example.com
use_fully_qualified_names = True
ldap_id_mapping = True
access_provider = ad
```

Since SSSD acts as an NSS provider and the default `nsswitch.conf` value for `passwd` is `files sss` in Fedora, we can query information about the `testuser` account using the usual `id` command, although in our configuration, we will need to specify its fully qualified name with the domain:

```
$ id testuser@ad.example.com
uid=1505601103(testuser@ad.example.com) gid=1505600513(domain users@
ad.example.com) groups=1505600513(domain users@ad.example.com)
```

It is also immediately possible to log in to the console by entering `testuser@ad.example.com` and the password we set on the controller earlier.

As you can see, setting up a domain controller that's compatible with Microsoft Active Directory using Samba is a simple and mostly automated procedure.

Summary

In this chapter, we learned how Linux-based systems implement authentication processes and user information lookup using PAM, NSS, and SSSD. We also learned about solutions for centralized authentication and demonstrated how to set up a domain controller that's compatible with Microsoft Active Directory using Samba software and join that domain from another Linux machine.

In the next chapter, we will learn about high-availability mechanisms in Linux.

Further reading

To learn more about the topics that were covered in this chapter, take a look at the following resources:

- SSSD documentation: `https://sssd.io/docs/introduction.html`
- Samba project: `https://www.samba.org/`
- FreeIPA project: `https://www.freeipa.org/`
- Kerberos: `https://en.wikipedia.org/wiki/Kerberos_(protocol)`
- LDAP: `https://en.wikipedia.org/wiki/Lightweight_Directory_Access_Protocol`
- Microsoft Active Directory: `https://en.wikipedia.org/wiki/Active_Directory`

13
High Availability

All computer hardware has limits regarding its performance and reliability, so systems that must process requests from large numbers of users without interruptions are always composed of multiple individual worker machines and dedicated load-balancing nodes that spread the load among those workers.

Linux includes functionality for load balancing and redundancy in the kernel, and multiple user-space daemons manage that built-in functionality and implement additional protocols and features.

In this chapter, we will learn about the following:

- Different types of redundancy and load balancing
- Link and network layer redundancy mechanisms in Linux
- Transport layer load balancing with **Linux Virtual Server** (**LVS**)
- Using Keepalived to share a virtual IP address between multiple nodes and automate LVS configuration
- Application layer load-balancing solutions, using HAProxy as an example

Types of redundancy and load balancing

Before we delve into specific high-availability features and their configuration, let's discuss possible types of redundancy and load balancing, their advantages, and their limitations.

First of all, we need to remember that the modern TCP/IP networking stack is *layered*. Multiple layering models include different numbers of layers but the idea is the same: protocols at the upper layer are unaware of the protocols at any lower levels and vice versa. The most commonly used models are the seven-layer **Open Systems Interconnection** (**OSI**) model and the four-level DoD model (developed by the United States Department of Defense). We have summarized them in the following table:

OSI model	DoD model	Purpose	Examples
Physical Data link	Link	Transmission of electrical/optical signals that represent bit streams	Ethernet, Wi-Fi
Network	Internet	Transmission of packets in segmented, routed networks	IPv4, IPv6
Transport	Transport	Reliable transmission of data segments (integrity checking, acknowledgment, congestion control, and more)	TCP, UDP, SSTP
Session Presentation Application	Application	Transmission of application-specific data	HTTP, SMTP, SSH

Table 13.1 — OSI and DoD network stack models

Since the network stack is layered, to make a network resistant to different types of failures, redundancy can and should be implemented at multiple levels. For example, connecting a single server to the network with two cables rather than one protects it from a single broken cable or a malfunctioning network card but will not protect the users from failures of the server software – in that case, the server will remain connected to the network but unable to serve any requests.

This problem can usually be solved by setting up multiple servers and introducing a dedicated load-balancer node to the network, which acts as an intermediary: it receives connections from users and distributes the load across all those servers.

Having a load balancer adds redundancy at the transport or application layer since the system remains able to serve requests, so long as at least one server is available. It also increases the total service capacity beyond the performance limit of a single server.

However, the load balancer itself becomes a single point of failure – if its software fails or it ends up disconnected from the network, the entire service becomes unavailable. Additionally, it becomes subject to the greatest network traffic load compared to any individual server. This makes link layer and network layer redundancy especially relevant. Finally, to make sure that requests that users send to the IP address of the load balancer are always accepted, the public address is often shared between multiple physical load-balancing servers in a cluster using the **First Hop Redundancy Protocol** (**FHRP**):

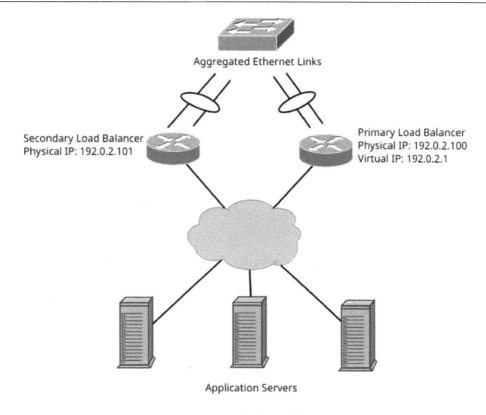

Figure 13.1 — Typical high-availability setup

Figure 13.1 shows a fully redundant setup with three application servers and two load balancers that are protected from cable or network card failures with aggregated Ethernet links.

Before we learn about different redundancy types and their implementations in Linux, we should review the terminology. Different protocols and technologies use different names for node roles, and some of them still use terminology that is inaccurate and may be offensive, but it's important to know it to understand what their documentation is talking about.

> **Notes on terminology**
>
> To describe redundant setups, we will use *active/standby* terminology by default: only one *active* node performs any work at a time and one or more additional *standby* nodes are waiting to take its place if it fails.
>
> A lot of older literature, including protocol standards, configuration files, and official documentation for high-availability solutions, may use *master/slave* terminology instead. That terminology is getting phased out by many projects due to its associations with human slavery and also because it is misleading since in most protocols, the active node does not have any control over standby nodes. We will use that terminology when we discuss protocols and software that still use it, for consistency with their documentation.

Link layer redundancy

Broken cables and Ethernet switch ports are quite common, especially in outdoor installations and industrial networks. In those situations, it is very useful to have more than one link layer connection. However, simply connecting two different network cards of a Linux machine to different ports of the same switch does not make them work as a single connection. The user needs to explicitly set up those two network cards so that they work together.

Luckily, Linux supports multiple ways to use several network cards together – both in active/standby and load-balancing configurations. Some of them do not require any support from the Ethernet switch and will work even with very basic unmanaged switches. Other modes require that the switch supports either the older EtherChannel protocol (designed by Cisco Systems) or the newer and vendor-neutral IEEE 802.3ad **Link Aggregation and Control Protocol** (**LACP**), and the ports must be configured explicitly to enable those protocols. We can summarize all these methods in the following table:

Type	Operation	Switch Requirements
`active-backup(1)`	One network card remains disabled while the other is up	None; will work for any switch (even unmanaged)
`802.3ad(4)`	Frames are balanced across all ports	Requires 803.3ad LACP support
`balance-xor(2)` and `broadcast(3)`		Requires EtherChannel support
`balance-tlb(5)` and `balance-alb(6)`		None

Table 13.2 — Link layer redundancy methods in Linux

The simplest mode is active-backup, which requires no special setup on the Ethernet switch and can even work with the simplest and cheapest unmanaged switches. Unlike modes such as 802.3ad LACP, it only provides active-standby redundancy rather than load balancing. Using the following command, you can join the `eth0` and `eth1` network interfaces to a single `bond0` interface using the active-backup method, on a system that uses NetworkManager for configuration:

```
$ sudo nmcli connection add type bond con-name bond0 ifname bond0
bond.options "mode=active-backup"
$ sudo nmcli connection add type ethernet slave-type bond con-name
bond0-port1 ifname eth0 master bond0
$ sudo nmcli connection add type ethernet slave-type bond con-name
bond0-port2 ifname eth1 master bond0
$ sudo nmcli connection up bond0
```

Now, if either `eth0` or `eth1` are physically disconnected from the switch, link layer connectivity will be preserved.

The cost of that configuration simplicity and low requirements for the Ethernet switch is wasted bandwidth. Whenever possible, high-performance servers should be connected to Ethernet networks using the current industry-standard 802.3ad LACP protocol, which allows them to benefit from the combined bandwidth of multiple links and also automatically exclude failed links to provide redundancy.

Network layer redundancy and load balancing

If a system has multiple independent connections to the internet or an internal network, it is possible to either provide a backup route or balance IP packets across multiple routes. However, in practice, network layer redundancy is only used by routers rather than hosts, and its simplest forms are only applicable to networks with public, globally routed addresses.

Suppose your Linux system is connected to two different routers, one with IPv4 address 192.0.2.1, and the other with 203.0.113.1. If you are fine with one connection being completely unused and acting purely as a standby, you can create two default routes with different *metrics* and assign a higher metric to the standby connection. The metric's value determines the route's priority, and if multiple routes with different metrics exist, the kernel will always use the route with the lowest metric. When that route disappears (for example, due to a network card going down), the kernel will switch to using the route with the next lowest metric of those still available.

For example, you can use the following commands if you want 192.0.2.1 to be the backup router:

```
$ sudo ip route add 0.0.0.0/0 via 192.0.2.1 metric 200
$ sudo ip route add 0.0.0.0/0 via 203.0.113.1 metric 100
```

The advantage of this method is that it is compatible with **Network Address Translation (NAT)** set up on the same system. If you want to create a load-balancing configuration instead, many more issues come into play because network layer load balancing is per-packet and unaware of any concept of a connection.

On the surface, the configuration for multi-path routes is quite simple. You can specify as many gateway addresses as you want and, optionally, assign weights to them to direct more traffic to faster links. For example, if you wanted twice as much traffic to flow through 203.0.113.1, you could achieve this with the following command:

```
$ sudo ip route add 0.0.0.0/0 nexthop via 192.0.2.1 weight 5 nexthop
via 203.0.113.1 weight 10
```

The problem is that this configuration, by itself, is incompatible with NAT because it will send packets that belong to the same TCP connection or a UDP stream to different gateways. If you have a publicly routed network, that is considered normal and even inevitable. However, if you only have a single external address from each provider and have to use NAT to map a private network to that single outgoing address, packets that belong to a single connection must always flow through the same gateway for the setup to work as expected. There are ways to set up per-connection load balancing using policy-based routing but that is outside the scope of this book. If you are interested, you can find more information in other sources, such as *Policy Routing With Linux*, by *Matthew G. Marsh*, which is freely available online.

Transport layer load balancing with LVS

The main disadvantage of all network layer mechanisms is that the network layer operates with individual packets and has no concept of connections. Many network services are connection-oriented so at the very least, all packets that belong to the same connection must always be sent to the same server. While the NAT implementation in Linux is smart enough to detect packets from the same connection, simple load balancing with one-to-many NAT is still too simplistic for many use cases. For example, it does not provide an easy way to track how many connections each server gets and cannot preferentially send new connections to the least loaded servers (that is, to servers that are handling the smallest number of existing connections at the moment).

To account for this use case, Linux includes the **LVS** framework and a tool for managing it – ipvsadm.

The key concepts of the LVS framework are *virtual servers* and *real servers*. Virtual servers are Linux machines that provide the public address of the service, accept connections to it, and then distribute those connections to multiple real servers. Real servers can run any OS and software and can be unaware of the virtual server's existence.

LVS is a flexible framework that provides multiple load-scheduling algorithms, load-balancing mechanisms, and configuration options, all with their advantages and disadvantages. Let's examine them in detail.

Scheduling algorithms

There are multiple ways to distribute the load between multiple servers, each with its advantages and disadvantages. We can summarize them in the following table:

Algorithm	Description
Round Robin (`rr`)	Distributes a connection across all servers equally.
Weighted Round Robin (`wrr`)	This is similar to Round Robin but allows you to send more connections to certain servers by assigning a higher weight value to them.
Least Connection (`lc`)	Preferentially sends new connections to the server with the least number of current connections.
Weighted Least Connection (`wlc`)	The default scheduling algorithm. This is similar to Least Connection but allows you to assign weights to servers.
Locality-Based Least-Connection (`lblc`)	Sends new connections with the same destination IP address to the same server, and switches to the next server if the first one is unavailable or overloaded.
Locality-Based Least-Connection with Replication (`lblcr`)	Sends new connections with the same destination IP address to the same server, if it is not overloaded. Otherwise, it sends them to the server with the least connections.
Destination Hashing (`dh`)	Creates a hash table that maps destination IP addresses to servers.
Source Hashing (`sh`)	Creates a hash table that maps source IP addresses to servers.
Shortest Expected Delay (`sed`)	Sends new connections to the server with the shortest expected delay.
Never Queue (`nq`)	Sends new connections to the first idle servers, and switches to Shortest Expected Delay if there are no idle servers.

Table 13.3 – LVS scheduling algorithms

The right choice of scheduling algorithm depends on the type of service; none of them is inherently better than others for all use cases. For example, Round Robin and Weighted Round Robin work best for services with short-lived connections, such as web servers that serve static pages or files (such as content delivery networks).

Services that use very long-lived, persistent connections, such as online game servers, can benefit from Least Connection algorithms instead. Using Round Robin methods for such services can be counter-productive because if new connections are relatively infrequent but resource consumption per connection is high, it can overload some of the servers or create a very unequal load distribution. Least Connection algorithms that keep track of the number of active connections to each server were designed to counter that problem.

Finally, if response latency is a big factor in the quality of service, the Shorted Expected Delay and Never Queue algorithms can improve it, while Round Robin and Least Connection do not take response time into account at all.

LVS load-balancing methods

First, we will examine the load-balancing methods that LVS provides. It supports three methods: direct routing, IP tunneling, and NAT. We will summarize the differences between them and their advantages and disadvantages in a table, then examine them in detail with configuration examples:

Mechanism	Implementation	Advantages	Disadvantages
Direct routing	Replaces the destination MAC address	Best performance; real servers send replies directly to clients	All servers must be on the same network It has difficulties with ARP
IP tunneling	Sends client requests encapsulated in a tunneling protocol	Real servers send replies directly to clients Real servers can be on any network	Real servers must support IPIP tunneling and must have tunnels to the virtual server The return packets may be rejected as spoofed
NAT	Creates NAT rules behind the scenes	Real servers don't need public addresses or any special configuration	Relatively resource-intensive All traffic goes through the virtual server The best method in practice despite its drawbacks

Table 13.4 – LVS load-balancing methods

Let's examine these load-balancing mechanisms in detail, starting with NAT.

NAT

NAT is the most practical load-balancing method of LVS because of two factors: real servers do not need to have publicly routable IP addresses and also do not need to be aware of the virtual server or specially configured to work with it.

The ability to use non-public internal addresses is especially important in IPv4 networks, considering the shortage of IPv4 addresses. The lack of special configuration requirements on the real servers also makes it possible to use any OS on them, and it simplifies the configuration process as well.

An additional advantage of this method is that TCP or UDP ports do not have to be the same on the virtual server and real servers since the virtual server performs translation anyway rather than forwarding unmodified IP packets.

We will set up the virtual server to listen for HTTP requests on `192.168.56.100:80` and forward those requests to port `8000` of real servers:

```
root@virtual-server# ipvsadm --add-service --tcp-service
192.168.56.100:80
root@virtual-server# ipvsadm --add-server --tcp-service
192.168.56.100:80 --real-server 10.20.30.2:8000 --masquerading
```

The first command creates a virtual server instance. The second command adds a real server to forward packets to – in our case, `10.20.30.2:8000`. Finally, the `--masquearding` (`-m`) option tells it to use the NAT method when sending connections to that server.

We used the long versions of all the `ipvsadm` command-line options here but the command could also be written in short form (with the Round-Robin scheduling algorithm specified, `-s rr`):

```
ipvsadm -A -t 192.168.56.100:80 -s rr.
```

Now, we can ensure that the virtual server is configured using the `ipvsadm -list` or `ipvsadm -l` command:

```
root@virtual-server# ipvsadm -list
IP Virtual Server version 1.2.1 (size=4096)
Prot LocalAddress:Port Scheduler Flags
  -> RemoteAddress:Port           Forward Weight ActiveConn InActConn
TCP  192.168.56.100:http rr
  -> 10.20.30.2:8000              Masq    1      0          0
```

Now, if we run `wget http://192.168.56.100:80` on the client machine and run a traffic capture on the real server, we will see the following output:

```
root@real-server# tcpdump -n -i eth1 -q tcp port 8000
tcpdump: verbose output suppressed, use -v or -vv for full protocol
decode
```

```
listening on eth1, link-type EN10MB (Ethernet), capture size 262144
bytes
23:56:16... IP 192.168.56.1.44320 > 10.20.30.2.8000: tcp 0
23:56:16... IP 10.20.30.2.8000 > 192.168.56.1.44320: tcp 0
23:56:16... IP 192.168.56.1.44320 > 10.20.30.2.8000: tcp 0
23:56:16... IP 192.168.56.1.44320 > 10.20.30.2.8000: tcp 129
```

On the virtual server, we will see a notably different output:

```
root@virtual-server# tcpdump -n -i eth0 -q tcp port 80
tcpdump: verbose output suppressed, use -v[v]... for full protocol
decode
listening on eth0, link-type EN10MB (Ethernet), snapshot length 262144
bytes
00:02:19... IP 192.168.56.1.32890 > 192.168.56.100.80: tcp 0
00:02:19... IP 192.168.56.100.80 > 192.168.56.1.32890: tcp 0
00:02:19... IP 192.168.56.1.32890 > 192.168.56.100.80: tcp 0
00:02:19... IP 192.168.56.1.32890 > 192.168.56.100.80: tcp 129
```

As you can see, the virtual server completely takes over the communication between the client and the real server. Theoretically, this is a disadvantage because it greatly increases the amount of traffic that flows through the virtual server. In practice, Linux network performance is pretty good even on modest hardware, so it is not a serious issue. Besides, application-specific load-balancing solutions also proxy all traffic through the server, so this is no worse than using a service such as HAProxy. Since packet forwarding and port/address translation happen in the kernel space, this method offers better performance than user-space load-balancing applications.

We will briefly examine the other load-balancing mechanisms, but for a variety of reasons, they are much less practical than NAT and normally need not be used.

Direct routing

To set up LVS for direct routing, we need to use the `--gatewaying` (`-g`) option when we add a real server:

```
root@virtual-server# ipvsadm --add-service --tcp-service
10.20.30.1:8000
root@virtual-server# ipvsadm --add-server --tcp-service
10.20.30.1:8000 --real-server 10.20.30.2 --gatewaying
```

With this setup, when the virtual server receives a request on `10.20.30.1:8000`, it will simply change the MAC address in that packet to the MAC address of the `10.20.30.2` real server and re-send it to the Ethernet network for the real server to receive. The real server will then reply directly to the client without creating any additional load on the virtual server.

While this method is theoretically the most performant and conceptually simplest, in reality, it places the hardest requirements on the real servers. The minimal requirement is that all real servers must be in the same broadcast network segment. The other requirement is that all real servers must also be able to respond to packets from the same virtual IP as the service IP, usually by having the virtual service IP assigned as an alias.

However, assigning the same IP address to multiple hosts creates an address conflict. To make the network function properly in the presence of an address conflict, all nodes except the virtual server must be made to ignore ARP requests for the virtual IP. This can be done, for example, with the `arptables` tool:

```
root@real-server# arptables -A IN -d 10.20.30.1 -j DROP
root@real-server# arptables -A OUT -d 10.20.30.1 -j mangle --mangle-
ip-s 10.20.30.2
```

To truly avoid this conflict and ensure that no real server answers an ARP request for the virtual IP, those rules need to be inserted before the address is assigned. This fact makes it difficult or even impossible to correctly configure real servers for this scheme using the usual network configuration methods, such as distribution-specific scripts or NetworkManager.

This fact makes this scheme impractical to implement in most networks, despite its theoretical advantages.

Tunneling

To set up a virtual server for tunneling, we need to use the `--ipip` (`-i`) option when we add a real server:

```
root@virtual-server# ipvsadm --add-service --tcp-service
192.168.56.100:8000
root@virtual-server# ipvsadm --add-server --tcp-service
192.168.56.100:8000 --real-server 10.20.30.2 --ipip
```

Then, we need to set up an IPIP tunnel on the real server so that it can handle incoming tunneled traffic from the virtual server and assign the virtual server IP to it:

```
root@real-server# ip tunnel add ipip1 mode ipip local 10.20.30.2
root@real-server# ip link set dev ipip1 up
root@real-server# ip address add 192.168.56.100/32 dev ipip1
```

Now, if we make an HTTP request to the virtual server and run a traffic capture on the real server, we will see incoming IPIP packets with requests for the virtual IP inside:

```
root@real-server# tcpdump -vvvv -n -i eth1
tcpdump: listening on eth1, link-type EN10MB (Ethernet), capture size
262144 bytes
01:06:05.545444 IP (tos 0x0, ttl 63, id 0, offset 0, flags [DF], proto
IPIP (4), length 80)
```

```
    10.20.30.1 > 10.20.30.2: IP (tos 0x0, ttl 63, id 44915, offset 0,
flags [DF], proto TCP (6), length 60)
    192.168.56.1.51886 > 192.168.56.106.8000: tcp 0
```

While this approach theoretically enables real servers to be in any network, it comes with several difficulties in practice. First, the real server OS must support IPIP tunneling. This can be a serious difficulty even with Linux systems if they run in containers and do not have permission to create tunnels, even if the host system kernel is built with IPIP support. Second, since replies are supposed to be sent directly to the client rather than back through the tunnel, this scheme falls apart in networks that take measures against source IP spoofing – as they should.

Saving and restoring LVS configurations

It is possible to export the current LVS configuration in a format that it can load from standard input:

```
root@virtual-server# ipvsadm --save
-A -t 192.168.56.100:http -s wlc
-a -t 192.168.56.100:http -r 10.20.30.2:8000 -m -w 1
```

You can save the output to a file and then feed it to ipvsadm -restore:

```
root@virtual-server# ipvsadm -save > lvs.conf
root@virtual-server# cat lvs.conf | ipvsadm --restore
```

However, in practice, it is better to automate LVS configuration with Keepalived or another user-space daemon, as we will learn later in this chapter.

Additional LVS options

In addition to scheduling algorithms and balancing between real servers, LVS offers a few additional features and options.

Connection persistence

By default, LVS balances connections from clients across all servers and does not match clients with specific servers. This approach works well for serving web pages over HTTP, for example. However, some services use long-lasting and stateful connections and would not work well without persistence. One extreme example is remote desktop connections: if such connections are balanced between multiple servers, sending a user to a different server after a disconnect will create a completely new session rather than get the user back to their already running applications.

To make LVS remember client-to-server mappings and send new connections from the same client to the same server, you need to specify --persistent and, optionally, specify a persistence timeout:

```
ipvsadm --add-service --tcp-service 192.168.56.100:80 --persistent 600
```

This preceding command creates a server that remembers client-to-server associations for 600 seconds.

Connection state synchronization

One notable feature of LVS is its connection state synchronization daemon. In that case, the word *daemon* is partially a misnomer since it is implemented in the kernel and is not a user-space process. Connection synchronization is unidirectional, with dedicated primary (master) and replica (backup) nodes.

There is no explicit peer configuration. Instead, connection states are sent to peers using IP multicast. It is possible to specify the network interface to use for synchronization messages:

```
root@first-virtual-server# ipvsadm --start-daemon=master --mcast-
interface=eth0
root@second-virtual-server# ipvsadm --start-daemon=backup --mcast-
interface=eth0
```

However, connection state synchronization by itself is useless, unless there's also a failover mechanism that allows you to transfer the virtual IP to the backup node if the primary load-balancer node fails.

In the next section, we will learn how to configure failover using the Keepalived daemon for VRRP.

Active/backup configurations and load balancing with Keepalived

A Linux server that is set up as a load balancer for multiple worker servers and keeps the service available, even if any of those workers fail. However, the load balancer itself becomes a single point of failure in that scheme, unless the administrator also takes care to provide a failover mechanism for multiple balancers.

The usual way to achieve failover is by using a floating *virtual IP address*. Suppose www.example.com is configured to point at 192.0.2.100. If you assign that address directly to a load-balancing server in a 192.0.2.0/24 network, it becomes a single point of failure. However, if you set up two servers with primary addresses from that network (say, 192.0.2.10 and 192.0.2.20), you can use a special failover protocol to allow two or more servers to decide which one will hold the virtual 192.0.2.100 address and automatically transfer it to a different server if the primary server fails.

The most popular protocol for that purpose is called **Virtual Router Redundancy Protocol (VRRP)**. Despite its name, machines that use VRRP do not have to be routers – even though it was originally implemented by router OSs, now, its use is much wider.

The most popular VRRP implementation for Linux is the Keepalived project. Apart from VRRP, it also implements a configuration frontend for LVS, so it is possible to write a configuration file for both failover and load balancing, without setting up LVS by hand with ipvsadm.

Installing Keepalived

Most Linux distributions have Keepalived in their repositories, so installing it is a straightforward process. On Fedora, RHEL, and its community derivatives such as Rocky Linux, you can install it using the following command:

```
sudo dnf install keepalived
```

On Debian, Ubuntu, and other distributions that use APT, run the following command:

```
sudo apt-get install keepalived
```

Now that we have installed Keepalived, let's look at the basics of the VRRP protocol.

Basics of the VRRP protocol operation

VRRP and similar protocols, such as the older **Hot Standby Router Protocol** (**HSRP**) and the community-developed **Common Address Redundancy Protocol** (**CARP**), are based on the idea of electing the primary node and continually checking its status by listening to its keepalive packets. Collectively, such protocols are known as **First Hop Redundancy Protocols** (**FHRPs**).

Initially, every node assumes that it may be the primary node and starts transmitting keepalive packets (named *advertisements* in the VRRP terminology) that include a unique identifier of the VRRP instance and a priority value. At the same time, they all start listening to incoming VRRP advertisement packets. If a node receives a packet with a priority value higher than its own, it assumes the backup role and stops transmitting keepalive packets. The node with the highest priority becomes the primary node and assigns the virtual address to itself.

The elected primary node keeps sending VRRP advertisement packets at regular intervals to signal that it is functional. Other nodes remain in the backup state, so long as they receive those packets. If the original primary node ceases to transmit VRRP packets, a new election is initiated.

If the original primary node reappears after a failure, there are two possible scenarios. By default, in the Keepalived implementation, the highest priority node will always preempt and the node that assumed its role during its downtime will go back to the backup state. This is usually a good idea because it keeps the primary router predictable under normal circumstances. However, preemption also causes an additional failover event that may lead to dropped connections and brief service interruptions. If such interruptions are undesirable, it is possible to disable preemption.

Configuring VRRP

Let's look at a simple example of VRRP configuration and then examine its options in detail:

```
vrrp_instance LoadBalancers {
    state BACKUP
    interface eth1
    virtual_router_id 100
    priority 100
    advert_int 1
    nopreempt
    virtual_ipaddress {
        10.20.30.100/24
    }
}
```

You will need to save that configuration to the Keepalived configuration file – typically, to `/etc/keepalived/keepalived.conf`.

The Keepalived configuration file may include one or more VRRP instances. Their names are purely informational and can be arbitrary, so long as they are unique within the configuration file.

The `state` option defines the initial state of the router. It is safe to specify `BACKUP` on all routers because they will elect the active router automatically, even if none of them has the `MASTER` state in its configuration.

VRRP instances are bound to network interfaces and exist in a single broadcast domain only, so we need to specify the network interface from which VRRP advertisements will originate. In that example, it is `interface eth1`.

The **Virtual Router ID (VRID)** defines the VRRP instance from the protocol's point of view. It is a number from 1 to 255, so there can be up to 254 distinct VRRP instances within the same broadcast network, so long as they use different VRIDs. There is no default value for that option and you cannot omit it. In our case, we used `virtual_router_id 100`.

The next two parameters are optional. The default VRRP router priority is 100 unless specified otherwise. If you want to specify router priorities manually, you can use numbers from 1 to 254 – priority numbers 0 and 255 are reserved and cannot be used. A higher priority value means that the router is more likely to be elected as an active (master) router.

The advertisement packet transmission interval (`advertise_interval`) defaults to one second and for most installations, it is a sensible setting. VRRP does not create much traffic, so there are no strong reasons to make the interval longer.

Finally, we specified a single virtual address, `10.20.30.100/24`. It is possible to specify up to 20 virtual addresses, separated by spaces. One thing to note is that all virtual addresses do not have to belong to the same network and do not have to be in the same network as the permanent, non-floating address of the network interface where the VRRP instance is running. It may even be possible to create redundant internet connections by assigning private IPv4 addresses to the WAN interfaces of two routers and setting up the public IPv4 addresses allocated by the internet service provider as virtual addresses.

Verifying VRRP's status

When you save the sample configuration to `/etc/keepalived/keepalived.conf` and start the process with `sudo systemctl start keepalived.service` (on Linux distributions with systemd), your server will become the active (master) node and assign the virtual address to its network interface, until and unless you add a second server with a higher priority to the same network.

The simplest way to verify this is to view IP addresses for the interface that we configured VRRP to run on:

```
$ ip address show eth1
3: eth1: <BROADCAST,MULTICAST,UP,LOWER_UP> mtu 1500 qdisc pfifo_fast
state UP group default qlen 1000
   link/ether 08:00:27:33:48:b8 brd ff:ff:ff:ff:ff:ff
   inet 10.20.30.1/24 brd 10.20.30.255 scope global eth1
   valid_lft forever preferred_lft forever
   inet 10.20.30.100/24 scope global secondary eth1
   valid_lft forever preferred_lft forever
   inet6 fe80::a00:27ff:fe33:48b8/64 scope link
   valid_lft forever preferred_lft forever
```

You can also use traffic capture tools such as `tcpdump` to verify that the server is indeed sending VRRP advertisement packets:

```
$ sudo tcpdump -i eth1
listening on eth1, link-type EN10MB (Ethernet), capture size 262144
bytes
04:38:54.038630 IP 10.20.30.1 > 224.0.0.18: VRRPv2, Advertisement,
vrid 100, prio 100, authtype none, intvl 1s, length 20
04:38:55.038799 IP 10.20.30.1 > 224.0.0.18: VRRPv2, Advertisement,
vrid 100, prio 100, authtype none, intvl 1s, length 20
04:38:56.039018 IP 10.20.30.1 > 224.0.0.18: VRRPv2, Advertisement,
vrid 100, prio 100, authtype none, intvl 1s, length 20
```

However, there is also a way to request VRRP's status data directly from Keepalived. Unlike some other services, Keepalived (as of its 2.2.7 release) does not include a socket interface or a command-line utility for interacting with it and uses POSIX signals to trigger state file creation. This is less convenient than a dedicated utility would be.

First, you need to look up the identifier (PID) of the Keepalived process. The best way to retrieve it is to read its PID file, most often located at `/run/keepalived.pid`.

Sending the `SIGUSR1` signal to the process with `kill -USR1 <PID>` will produce a data file at `/tmp/keepalived.data`. This file contains multiple sections, and the section of immediate interest for us to find out the status of our VRRP instance is named **VRRP Topology**:

```
$ cat /run/keepalived/keepalived.pid
3241

$ sudo kill -USR1 3241

$ sudo cat /etc/keepalived.data
...
------< VRRP Topology >------
  VRRP Instance = LoadBalancers
   VRRP Version = 2
   State = MASTER
   Flags: none
   Wantstate = MASTER
   Last transition = ...
   Interface = eth1
   Using src_ip = 10.20.30.1
   Multicast address 224.0.0.18
   ...
   Virtual Router ID = 100
   Priority = 100
   ...
   Preempt = enabled
   Promote_secondaries = disabled
   Authentication type = none
   Virtual IP (1):
     10.20.30.100/24 dev eth1 scope global set
   ...
```

It is also possible to request a statistics file (`/tmp/keepalived.stats`) by sending the Keepalived process the `SIGUSR2` signal instead:

```
$ sudo kill -USR1 $(cat /run/keepalived.pid)

$ sudo cat /etc/keepalived.stats
VRRP Instance: LoadBalancers
  Advertisements:
    Received: 0
```

```
   Sent: 112
Became master: 1
Released master: 0
Packet Errors:
   Length: 0
   TTL: 0
   Invalid Type: 0
   Advertisement Interval: 0
   Address List: 0
Authentication Errors:
   Invalid Type: 0
   Type Mismatch: 0
   Failure: 0
Priority Zero:
   Received: 0
   Sent: 0
```

While the information method is somewhat unwieldy at the moment, you can glean a lot of information about your VRRP instances from those data files.

Configuring virtual servers

As we already said, Keepalived can also create and maintain LVS configurations. The advantage over configuring LVS manually is that Keepalived is easy to start at boot time since it always comes with service management integration (typically, a systemd unit), while LVS is a kernel component that does not have a configuration persistence mechanism. Additionally, Keepalived can perform health checks and reconfigure the LVS subsystem when servers fail.

For demonstration purposes, let's consider a minimal load-balancing configuration with a Weighted Round Robin balancing algorithm, NAT as the load-balancing method, and two real servers with equal weights:

```
global_defs {
  lvs_id WEB_SERVERS
}
virtual_server 192.168.56.1 80 {
    ! Weighted Round Robin
    lb_algo wrr
    lb_kind NAT
    protocol TCP

    ! Where to send requests if all servers fail
    sorry_server 192.168.56.250 80
    real_server 192.168.56.101 80 {
```

```
        weight 1
    }
    real_server 192.168.56.102 80 {
        weight 1
    }
}
```

Every load-balancing algorithm that we discussed in the *Transport layer load balancing with LVS* section can be specified in the `lb_algo` option, so it could be `lb_algo wlc` (Weighted Least Connection), for example.

If you save that configuration to `/etc/keepalived/keepalived.conf` and restart the daemon with `systemctl restart keepalived`, you can verify that it created an LVS configuration:

```
$ sudo systemctl restart keepalived
$ sudo ipvsadm
IP Virtual Server version 1.2.1 (size=4096)
Prot LocalAddress:Port Scheduler Flags
  -> RemoteAddress:Port           Forward Weight ActiveConn InActConn
TCP  server:http wrr
  -> 192.168.56.101:http          Masq    1       0         0
  -> 192.168.56.102:http          Masq    1       0         0
```

Now that we know how to make a basic virtual server configuration, let's learn how to monitor the status of real servers and exclude them if they fail.

Server health tracking

LVS by itself is purely a load-balancing solution and it does not include a server health monitoring component. However, in real-world installations, prompt exclusion of servers that are not functioning correctly or are under scheduled maintenance is an essential task, since directing user requests to non-functional servers defeats the purpose of a high-availability configuration. Keepalived includes monitoring capabilities so that it can detect and remove servers that fail health checks.

Health checks are configured separately for each real server, although in most real-world installations, they should logically be the same for all servers, and using different health check settings for different servers is usually a bad idea.

TCP and UDP connection checks

The simplest but the least specific health check type is a simple connection check. It exists in two variants – UDP_CHECK and TCP_CHECK for UDP and TCP protocols, respectively. Here is a configuration example for that check type:

```
real_server 192.168.56.101 80 {
    weight 1
    TCP_CHECK {
        connect_timeout 3
        retry 3
        delay_before_retry 2
    }
}
```

As you can see, there is no need to specify the TCP port for connection checks explicitly: Keepalived will use the port specified in the server address configuration (port 80 in this case).

When you start Keepalived with that configuration, it will activate the health-checking subsystem and begin connection checks. If there is no running web server on 192.168.56.101 listening on port 80, Keepalived will remove that server from the LVS configuration once its check fails three times (as defined by the retry option). You will see the following in the system log (which you can view, for example, with sudo journalctl -u keepalived):

```
Keepalived_healthcheckers: Activating healthchecker for service
[192.168.56.101]:tcp:80 for VS [192.168.56.1]:tcp:80
Keepalived: Startup complete
Keepalived_healthcheckers: TCP_CHECK on service
[192.168.56.101]:tcp:80 failed.
Keepalived_healthcheckers: Removing service [192.168.56.101]:tcp:80
from VS [192.168.56.1]:tcp:80
```

The advantage of this simple TCP check is that it works for any TCP-based service, no matter what its application layer protocol is: you can use it for web applications, as well as SMTP servers or any custom protocols. However, the fact that a server responds to TCP connections by itself does not always mean that it is also functioning correctly. For example, a web server may respond to TCP connections but reply to every request with a **500 Internal Server Error** result.

If you want perfect, fine-grained control over the check logic, Keepalived gives you that option in the form of the MISC_CHECK method.

Misc (arbitrary script) check

The most universal check is MISC_CHECK, which does not have any built-in checking logic and relies on an external script instead. For example, this is how you can make Keepalived execute the /tmp/my_check.sh script and consider the server unavailable if that script returns a non-zero exit code:

```
real_server 192.168.56.101 80 {
    MISC_CHECK {
        misc_path "/tmp/my_check.sh"
        misc_timeout 5
        user nobody
    }
}
```

With this type of health check, you can monitor any kind of server, although the disadvantage is that you have to implement all the checking logic yourself in a script.

HTTP and HTTPS checks

While MISC_CHECK gives you total control, it is also overkill in most cases.

As a compromise between specificity and flexibility, you can also use protocol-specific checks. For example, there is the HTTP_GET check, which makes an HTTP request to a URL and can check the hash sum of the response, or its HTTPS equivalent named SSL_CHECK.

For example, suppose you want to serve a simple static page. In that case, you can calculate an MD5 hash sum from that page by hand using the md5sum command:

```
$ cat index.html
<!DOCTYPE html>
<html>
  <body>
    <p>hello world</p>
  </body>
</html>
$ md5sum index.html
fecf605e44acaaab933e7b509dbde185  index.html
```

To calculate the expected hash sum of a dynamically generated page, you can use the genhash utility that comes with Keepalived. If you run it with --verbose, it will show you detailed information about the HTTP request it performs:

```
$ genhash --verbose --server 192.168.56.101 --port 80 --url /index.
html
----------[    HTTP Header Buffer    ]----------
0000    48 54 54 50 2f 31 2e 30 - 20 32 30 30 20 4f 4b 0d    HTTP/1.0 200
```

```
OK.
...
----------[ HTTP Header Ascii Buffer ]----------
HTTP/1.0 200 OK
...
----------[        HTML Buffer        ]----------
0000   3c 21 44 4f 43 54 59 50 - 45 20 68 74 6d 6c 3e 0a    <!DOCTYPE
html>.

...


---------[     HTML hash resulting     ]---------
0000   fe cf 60 5e 44 ac aa ab - 93 3e 7b 50 9d bd e1
85    ..`^D....>{P....
--------[ HTML hash final resulting ]--------
fecf605e44acaaab933e7b509dbde185
Global response time for [/index.html] = 2468 usecs
```

However, it only calculates the hash sum of the HTTP response body rather than the complete response with headers, so you do not have to use it – you can retrieve the response body with any other HTTP request utility if you prefer.

Once you have the expected response hash sum, you can configure the HTTP_GET check to periodically perform a request and check its response body against the given MD5 sum:

```
real_server 192.168.56.101 80 {
    weight 1
    HTTP_GET {
        url {
            path /index.html
            digest fecf605e44acaaab933e7b509dbde185
        }
        connect_timeout 3
        retry 3
        delay_before_retry 2
    }
}
```

Since normal, user-visible pages can change at any time, it is better to create a special page whose content stays constant if you want to use the hash sum check. Otherwise, the hash sum will change when its content changes, and the check will start failing.

Email notifications

It is also possible to configure Keepalived to send email notifications to one or more addresses when any status changes occur – that is, when VRRP transitions from master to backup or the other way around, or when real servers become unavailable and fail checks, or pass checks that were failing earlier and are added back to the LVS configuration in the kernel.

Here is a configuration example:

```
global_defs {
    notification_email {
        admin@example.com
        webmaster@example.com
    }
    notification_email_from keepalived@example.com
    smtp_server 203.0.113.100
    smtp_connect_timeout 30
}
```

Unfortunately, there is no support for SMTP authentication, so if you choose to use the built-in email notification mechanism, you need to configure a server as an open relay and take appropriate measures to ensure that only the servers running Keepalived can send messages through it – for example, by limiting access to it to your private network using firewall rules.

Application layer load balancing

LVS is a flexible framework for load balancing and the fact that it is implemented within the kernel makes it a high-performance solution since it does not require context switches and data transfer between user-space programs and the kernel. The fact that it works at the TCP or UDP protocol level also makes it application-agnostic and allows you to use it with any application service.

However, its lack of application protocol awareness is also its greatest weakness because it means that it cannot perform any protocol-specific optimizations. For example, one obvious way to improve performance for applications that may return the same reply to multiple users is to cache replies. LVS operates with TCP connections or UDP streams, so it has no way to know what a request or a reply looks like in any application layer protocol – it simply does not inspect TCP or UDP payloads at all.

Additionally, many modern application layer protocols are encrypted, so it is impossible to look inside the payload of a connection that the server does not initiate or terminate.

There are more potential disadvantages to forwarding connections directly from users to real servers. For example, it exposes servers to TCP-based attacks such as SYN flood and requires appropriate security measures on all servers or a dedicated firewall setup at the entry point.

One way to solve these issues is to use a user-space daemon that implements the protocol of the service you are running, terminates TCP connections, and forwards application layer protocol requests to target servers.

Since most applications in the world are currently web applications, most such solutions target HTTP and HTTPS. They provide in-memory response caching to speed up replies, terminate SSL connections, and manage certificates, and can optionally provide security features as well. HAProxy and Varnish are prominent examples of web application load-balancing servers, although there are other solutions for that purpose as well.

There are also solutions for other protocols that include high availability and load balancing. For example, OpenSIPS and FreeSWITCH can provide load balancing for **Voice over Internet Protocol** (**VoIP**) calls made using the SIP protocol. Such solutions are beyond the scope of this book, however. We will take a quick look at HAProxy as one of the most popular high-availability solutions for web applications.

Web application load balancing with HAProxy

HAProxy configuration is a large subject since it includes a lot of functionality. We will examine a simple configuration example to get a sense of its capabilities:

```
frontend main
    bind *:80
    acl url_static path_beg -i /static /images /javascript /
stylesheets
    acl url_static path_end -i .jpg .gif .png .css .js
    use_backend static           if url_static
    default_backend              app
backend static
    balance       roundrobin
    server static 192.168.56.200:80 check
backend app
    balance       roundrobin
    server  srv1 192.168.56.101:5000 check
    server  srv2 192.168.56.102:5000 check
```

As you can see, at its core, any HAProxy configuration maps frontends (that is, load-balancing instances) with backends – sets of actual application servers.

In this case, a single frontend is mapped to two backends: a single server specially for serving static files and two application servers. This is only possible for HAProxy because it handles HTTP requests itself, sends new requests to its backends, and prepares a reply to the user, instead of simply balancing connections.

Summary

In this chapter, we learned about the concepts of high availability: redundancy, failover, and load balancing. We also learned how to configure link-layer redundancy by creating bonding interfaces, as well as how to set up redundant routes at the network layer. To ensure transport layer redundancy, we learned how to configure the LVS subsystem by hand with `ipvsadm` or using Keepalived and also learned how to provide failover for load-balancing nodes using VRRP. Finally, we took a brief look at HAProxy as an application layer load-balancing solution for web servers.

In the next chapter, we will learn about managing Linux systems with configuration automation tools.

Further reading

To learn more about the topics that were covered in this chapter, take a look at the following resources:

- *Policy Routing With Linux*, by Matthew G. Marsh: `https://web.archive.org/web/20230322065520/http://www.policyrouting.org/PolicyRoutingBook/ONLINE/TOC.html`

- Keepalived documentation: `https://keepalived.readthedocs.io/en/latest/`

- HAProxy: `http://www.haproxy.org/`

14
Automation with Chef

In today's technology-driven world, managing and maintaining infrastructure at scale has become increasingly complex. System administrators often face challenges in deploying and configuring numerous servers, ensuring consistency across environments, and efficiently managing updates and changes. Manual configuration processes are time-consuming, error-prone, and difficult to scale. To address these issues, automation has emerged as a crucial solution. Among the various automation tools available, Chef stands out as a powerful configuration management system that streamlines infrastructure provisioning and management.

The purpose of this chapter is to provide a comprehensive overview of automating infrastructure with Chef in the Linux environment. It aims to explore the various components of Chef and their roles, examine the benefits of using Chef for infrastructure automation, discuss real-world use cases, and analyze Chef's strengths and limitations. By the end of this chapter, you will have gained a clear understanding of how Chef can revolutionize infrastructure management in Linux-based systems.

In this chapter, we will cover the following topics:

- Overview of infrastructure automation
- Introduction to Chef
- Chef server
- Chef workstation

Overview of infrastructure automation

Infrastructure automation is essential to overcome the challenges associated with managing complex and dynamic environments. By automating repetitive tasks, administrators can reduce human error, increase efficiency, and ensure consistent configurations across all servers. Automation also enables faster deployment, improves scalability, and enhances system security. In Linux environments, where flexibility and customization are paramount, infrastructure automation becomes even more crucial.

Benefits of automation in Linux

Automating infrastructure in Linux offers several benefits. Firstly, it allows rapid and consistent server provisioning, reducing the time required for manual configuration. Secondly, automation ensures adherence to standard configurations, minimizing inconsistencies and facilitating easier troubleshooting. Additionally, automation improves scalability by enabling the quick addition or removal of servers based on demand. Finally, automation enhances security by enforcing consistent security policies and facilitating timely updates and patch management. Now that we have understood the benefits of automation, let's go ahead and see how automation works in the following section.

Introduction to Chef

At its core, Chef follows an **Infrastructure as Code** (**IaC**) approach, where the desired state of a system or server is defined using code. This code, written in a **domain-specific language** (**DSL**) called the Chef DSL, describes the configuration, installation, and management of various components, applications, and services on a system.

What is Chef?

Chef is an open source configuration management tool that allows administrators to define and automate IaC. It follows a declarative approach, where administrators specify the desired state of the system, and Chef ensures the system conforms to that state. Chef provides a platform-agnostic solution, enabling automation across various OSs, including Linux. It is based on a client-server architecture and utilizes a DSL called Ruby.

Key features of Chef

Chef offers a set of powerful features that facilitate infrastructure automation. These features include the following:

- **Infrastructure as Code**: Chef treats infrastructure configuration as code, allowing administrators to version control, test, and deploy infrastructure changes with ease
- **Idempotent operations**: Chef ensures idempotency by running configuration recipes only when necessary, which eliminates the risk of unintended changes
- **Resource abstraction**: Chef abstracts system resources (such as files, services, and packages) into manageable components, simplifying configuration management
- **Testability**: Chef supports test-driven infrastructure development, enabling administrators to validate and test their configurations before deployment

Using automation saves a lot of time and reduces the risk of human errors.

Overview of Chef's architecture

Chef follows a client-server architecture. The key components of Chef's architecture are as follows:

- **Chef server**: The central component that stores and manages the configuration data, policies, and cookbooks. It acts as the authoritative source of truth for the desired system state.

- **Chef workstation**: The administrative machine where administrators author and test cookbooks and manage the infrastructure. It hosts the development environment and Chef client tools.

- **Chef nodes**: The target machines that are managed and configured by Chef. Each node runs a Chef client, which communicates with the Chef server to retrieve configuration instructions and apply them to the node.

These components will be covered in more detail in the following sections.

Chef server

The Chef server is the heart of the Chef infrastructure. It acts as the central repository for storing configuration data, cookbooks, policies, and node information. The server provides a web interface and API to interact with Chef resources. Administrators use the Chef server to manage nodes, roles, environments, and data bags, and to distribute cookbooks to nodes.

The Chef server utilizes a push/pull mechanism to manage the configuration of nodes (servers) in a system. This mechanism allows administrators to define desired states for nodes and enforce those states on the nodes:

- **Push mechanism**: In the push mechanism, the Chef server actively pushes the configuration updates and recipes to the nodes. When administrators make changes to the configurations or define new recipes, they upload those changes to the Chef server. The Chef server then identifies the target nodes and pushes the updated configurations to them. This process can be initiated manually or through automated processes.

- **Pull mechanism**: In the pull mechanism, nodes are configured to check the Chef server periodically for updates. Nodes will request and pull their configurations from the Chef server at regular intervals. When a node checks for updates, it compares its current state with the desired state specified on the Chef server. If there are any differences, the node pulls the necessary configurations and updates itself accordingly.

Chef server components

The key components of the Chef server include the following:

- **Data store**: The data store is where the Chef server stores the metadata and configuration information of nodes, cookbooks, roles, environments, and data bags. It utilizes a database (such as PostgreSQL) to store this information.

- **Chef server API**: The Chef server API provides a RESTful interface for managing and interacting with Chef resources programmatically. It allows administrators to query, modify, and delete resources.

- **User and authentication management**: The Chef server manages user accounts and authentication mechanisms. It provides **role-based access control (RBAC)** to control user permissions and restrict access to sensitive information.

The Chef server plays a vital role in enabling efficient infrastructure automation and configuration management, supporting scalability, consistency, and security in large-scale deployments. By providing a centralized hub for managing configurations and cookbooks, it simplifies the process of deploying and maintaining complex systems while promoting collaboration and best practices among teams of system administrators and developers.

Cookbooks and recipes

Cookbooks are the fundamental building blocks of Chef. They contain recipes, attributes, templates, and other files required to configure and manage specific components of the infrastructure. The Chef server acts as a repository for storing and managing cookbooks. Administrators can upload, version, and distribute cookbooks to the nodes via the Chef server. Cookbooks are organized into logical groups and are defined using a directory structure.

Using Chef, system administrators can automate tasks such as package installation, configuration file management, service management, and more. Chef's declarative nature allows for easy scalability and reproducibility, making it ideal for managing complex infrastructures.

Chef workstation

The Chef workstation serves as the administrative machine where administrators develop, test, and manage the Chef infrastructure. To set up a Chef workstation, administrators install the **Chef Development Kit (ChefDK)** on their local machine. The ChefDK includes all the tools, libraries, and dependencies required for cookbook development and management.

Development workflow

The Chef workstation provides a development environment where administrators author and test cookbooks before deploying them to nodes. Administrators use text editors or **integrated development environments** (**IDEs**) to write cookbooks using the Ruby-based Chef DSL. The workstation also includes tools such as Test Kitchen, which allows for cookbook testing in various virtualized or containerized environments.

Managing environments and roles

In Chef, environments and roles play a crucial role in managing infrastructure configuration. Administrators define environments to represent different stages (such as development, testing, and production) of the infrastructure. Environments enable administrators to set environment-specific attributes and control the cookbook versions deployed to each environment. Roles, on the other hand, define the desired state of a node based on its purpose or function. They group attributes and recipes required for a specific role and can be applied to multiple nodes.

Here's an example of a Chef environment JSON file:

```
{
  "name": "my_environment",
  "description": "Sample environment for my application",
  "cookbook_versions": {
    "my_cookbook": "= 1.0.0",
    "another_cookbook": "= 2.3.1"
  },
  "default_attributes": {
    "my_app": {
      "port": 8080,
      "debug_mode": false
    },
    "another_app": {
      "enabled": true
    }
  },
  "override_attributes": {
    "my_app": {
      "port": 8888
    }
  },
  "json_class": "Chef::Environment",
  "chef_type": "environment"
}
```

In this example, we define an environment named `my_environment`. It has a `description` field that provides a brief description of the environment.

The `cookbook_versions` section specifies the desired versions of cookbooks in the environment. In this case, `my_cookbook` is set to version `1.0.0`, and `another_cookbook` is set to version `2.3.1`.

The `default_attributes` section contains default attribute values for the environment. It sets the `port` attribute of `my_app` to `8080` and the `debug_mode` attribute to `false`. It also sets the `enabled` attribute of `another_app` to `true`.

The `override_attributes` section allows you to override specific attribute values from the cookbooks. In this example, it sets the `port` attribute of `my_app` to `8888`, overriding the default value.

The `json_class` and `chef_type` fields are required and specify the class and type of the environment, respectively.

To create or update an environment using this JSON file, you can use the `knife` command-line tool or Chef API. For example, with **knife**, you can run the following command:

```
knife environment from file my_environment.json
```

Make sure to replace `my_environment.json` with the actual filename and adjust the contents of the JSON file according to your specific environment configuration.

Now, let's look at a JSON template for role configuration:

```
{
  "name": "webserver",
  "description": "Role for web server nodes",
  "json_class": "Chef::Role",
  "chef_type": "role",
  "run_list": [
    "recipe[my_cookbook::default]",
    "recipe[another_cookbook::setup]"
  ],
  "default_attributes": {
    "my_cookbook": {
      "port": 8080,
      "debug_mode": false
    },
    "another_cookbook": {
      "config_file": "/etc/another_cookbook.conf"
    }
  },
  "override_attributes": {
    "my_cookbook": {
```

```
      "port": 8888
    }
  },
  "env_run_lists": {
    "production": [
      "recipe[my_cookbook::production]"
    ],
    "development": [
      "recipe[my_cookbook::development]"
    ]
  }
}
```

In this example, we defined a role named `webserver`. It has a `description` field that provides a brief description of the role.

The `run_list` section specifies the list of recipes to be included in the role's run list. In this case, it includes the `default` recipe from the `my_cookbook` cookbook and the `setup` recipe from the `another_cookbook` cookbook.

The `default_attributes` section contains default attribute values for the role. It sets the `port` attribute of `my_cookbook` to `8080` and the `debug_mode` attribute to `false`. It also sets the `config_file` attribute of `another_cookbook` to `/etc/another_cookbook.conf`.

The `override_attributes` section allows you to override specific attribute values from the cookbooks. In this example, it sets the `port` attribute of `my_cookbook` to `8888`, overriding the default value.

The `env_run_lists` section specifies different run lists for different environments. In this case, it includes the `production` run list, which includes the `production` recipe from the `my_cookbook` cookbook, and the `development` run list, which includes the `development` recipe from the `my_cookbook` cookbook.

The `json_class` and `chef_type` fields are required and specify the class and type of the role, respectively.

To create or update a role using this JSON file, you can use the knife command-line tool or Chef API. For example, with `knife`, you can run the following command:

```
knife role from file webserver.json
```

Make sure to replace `webserver.json` with the actual filename and adjust the contents of the JSON file according to your specific role configuration.

Roles and environments are very useful features of Chef as they can make life easier; for example, you can just mention one role that can include dozens of recipes rather than mention each of them.

Chef nodes

Chef nodes are the target machines that are managed and configured by Chef. Each node has a unique identity and requires the Chef client to be installed. Administrators define attributes and roles for each node in the Chef server, which determine how the node is configured and what recipes are applied.

Node registration

To join a node to the Chef infrastructure, administrators bootstrap the node by installing the Chef client and registering it with the Chef server. During the bootstrap process, the node generates a client key that allows it to authenticate with the Chef server securely.

Communication with the Chef server

Once registered, the Chef client on the node communicates with the Chef server to retrieve its configuration instructions. The client periodically converges with the server to ensure that the node's state aligns with the desired state defined by the Chef server. The Chef client applies the necessary changes to the node's configuration to achieve convergence.

A Chef node communicates with the Chef server using a secure and authenticated process. The communication between the Chef client (running on the node) and the Chef server is based on HTTPS and relies on cryptographic keys and certificates for authentication. Here's how the communication process works and what configuration is needed:

- **Client configuration**: The Chef client on the node needs to be properly configured with essential settings. The main configuration file for the Chef client is typically located at `/etc/chef/client.rb` (on Linux systems) or an equivalent location.

- **Client identity and validation**: The Chef client needs to have a unique identity associated with it, which is represented by a client name (usually the node's hostname) and a client key.

  A client key is a private key generated for the client, and it must be securely stored on the node. This key is used to authenticate the Chef client when communicating with the Chef server.

- **Chef server URL**: The Chef client needs to know the URL of the Chef server to establish a connection. This URL is specified in the client configuration file.

- **Validation key**: The Chef server issues a validation key (also known as a *validator key*) that is used by new nodes to register themselves with the Chef server.

  This validation key must be placed on the node, usually in a file named `validation.pem` (again, located in `/etc/chef/` on Linux systems).

- **Node registration**: When a new node (with the Chef client installed) comes online, it uses the validation key to register itself with the Chef server.

During the registration process, the node provides its client name and the client key to authenticate itself.

By setting up the correct configuration and ensuring the appropriate cryptographic keys and certificates are in place, a Chef node can securely communicate with the Chef server, fetch the latest configuration data, and perform Chef runs to converge its configuration to the desired state.

Cookbook development

Cookbooks are the core units of configuration management in Chef. They consist of a collection of related recipes, attributes, templates, and other files required to configure and manage specific aspects of the infrastructure. Cookbooks are organized into directories, each representing a specific component or service to be managed.

Cookbook structure and components

A cookbook follows a specific directory structure that includes the following components:

- **Recipes**: Recipes are the primary building blocks of a cookbook. They define the steps and resources required to configure a specific component or service. Recipes can include other recipes and leverage attributes to define the desired state.

- **Attributes**: Attributes allow administrators to define variables that customize the behavior of recipes. They can be used to specify package versions, file paths, service configurations, and other parameters.

- **Templates**: Templates are used to generate configuration files dynamically. They can contain **Embedded Ruby (ERB)** code to inject attributes or dynamically generate content.

- **Files**: Cookbooks can include additional files required for configuration, such as scripts, certificates, or binaries.

A collection of recipes, templates, and other resources are grouped to manage a specific set of related configurations. Cookbooks provide a modular and reusable way to manage various aspects of a system.

Writing recipes and resources

As mentioned earlier, recipes define the steps and resources required to configure a specific component or service. Resources represent individual elements of the system, such as packages, files, services, or users. Administrators use resource types provided by Chef, such as `package`, `file`, `service`, `template`, and so on, to define the desired state of each resource. By specifying the desired state, administrators allow Chef to converge the node's configuration to match that state.

Here's an example of a simple Chef recipe to install a package:

```
# Recipe: install_package
package 'my_package' do
  action :install
end
```

In this example, we have a recipe named `install_package`. The `package` resource is used to manage the installation of a package named `my_package`. By default, the `action` attribute of the `package` resource is set to `:install`, which instructs Chef to install the package. If the package is already installed, it will just move forward.

To use this recipe, you can create a cookbook and place this recipe in the appropriate recipe file (for example, `recipes/default.rb`). Make sure that the package you want to install is available in the package manager repository for your system.

Once the recipe has been set up, you can run Chef on a node with the cookbook to trigger the installation of the specified package.

An example of a resource template is as follows:

```
# Resource definition
resource_name :my_resourceproperty :name, String, name_property: true
property :port, Integer, default: 8080
property :enabled, [true, false], default: true

# Actions
default_action :create

action :create do
  template "/etc/myapp/#{new_resource.name}.conf" do
    source 'myapp.conf.erb'
    variables port: new_resource.port, enabled: new_resource.enabled
    owner 'root'
    group 'root'
    mode '0644'
    action :create
  end
end

action :delete do
  file "/etc/myapp/#{new_resource.name}.conf" do
    action :delete
  end
end
```

In this example, we defined a resource called my_resource. It has three properties: name (String), port (Integer), and enabled (Boolean). The name property is marked as the name property using name_property: true. The port property has a default value of 8080, and the enabled property has a default value of true.

The default action for this resource is set to :create. Inside the :create action, we use the template resource to generate a configuration file for our application. The template source is specified as myapp.conf.erb, which means it will use the corresponding ERB template file. We pass the port and enabled variables to the template using the variables attribute. The template file will be created with the owner and group set to root, and the file permissions set to 0644.

The :delete action uses the file resource to delete the configuration file.

You can customize this template according to your specific requirements. Remember to replace myapp.conf.erb with your actual template filename and adjust the paths and permissions as needed.

Managing infrastructure with Chef

Node bootstrapping is the process of preparing a target machine to be managed by Chef. It involves installing the Chef client and registering the node with the Chef server. Bootstrapping can be done manually or automated using tools like knife, which is a command-line utility provided by Chef.

Configuration management

Chef enables administrators to define and manage the configuration of infrastructure components using cookbooks and recipes. Administrators can specify the desired state of each resource, and Chef ensures that the node's configuration converges to that state. Configuration management includes tasks such as installing packages, managing files and directories, configuring services, and setting up networking.

Chef client-server interaction

The Chef client on each node periodically converges with the Chef server to ensure the node's configuration matches the desired state defined in the Chef server. During convergence, the client retrieves updated cookbooks, attributes, and recipes from the server. It compares the current state of the node with the desired state and makes necessary changes to achieve convergence.

Reporting and monitoring

Chef provides reporting and monitoring capabilities to track the status of the infrastructure and Chef client runs. Administrators can view reports on cookbook versions, node status, and convergence details. This information helps in troubleshooting, auditing, and ensuring compliance with configuration policies:

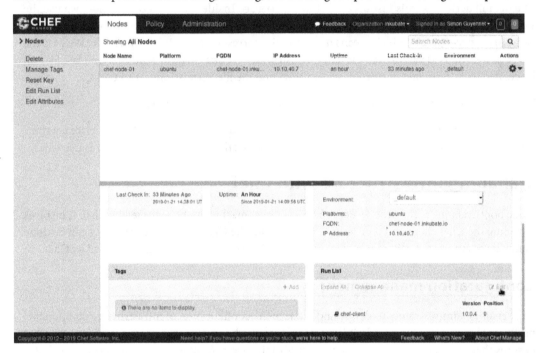

Figure 14.1 – Chef Dashboard (Source: https://docs.chef.io/manage/)

As you can see, it is very easy to monitor and manage everything from one dashboard.

Data synchronization

Chef enables data synchronization between the server and nodes through the use of data bags and encrypted data bags. Data bags are JSON data structures that store arbitrary data used by cookbooks. They allow for sharing data across nodes and providing configuration-specific information. Encrypted data bags provide an additional layer of security by encrypting sensitive data before storing it in the Chef server.

Benefits of automating infrastructure with Chef in Linux

Automating infrastructure with Chef in Linux significantly improves operational efficiency. It reduces manual configuration efforts, enabling administrators to deploy and manage infrastructure at scale. Chef's idempotent operations ensure that configurations are applied only when necessary, saving time and reducing errors.

Consistency and scalability

Chef ensures consistent configurations across distributed environments. By defining the desired state in cookbooks and recipes, administrators can easily replicate configurations across nodes, ensuring uniformity and eliminating configuration drift. Chef's scalability allows for easy addition or removal of nodes, accommodating dynamic infrastructure requirements.

Reduced human error

Automating infrastructure with Chef minimizes human error by removing manual intervention. Chef's declarative approach ensures that configurations are applied consistently and according to predefined rules. This reduces the risk of misconfigurations and enhances system stability and reliability.

Enhanced security

Chef enhances security by providing a centralized mechanism for managing configuration policies and enforcing consistent security practices. It allows administrators to define and distribute security-related configurations, such as firewall rules, user permissions, and access controls, across nodes. Regular updates and patch management can also be automated, ensuring the timely application of security fixes.

Challenges and best practices

Implementing Chef automation may pose certain challenges, such as the learning curve associated with the Chef DSL, managing complex dependencies, and handling infrastructure drift. To overcome these challenges, administrators can invest in training and documentation, adopt version control practices, and use Chef features such as environments and roles effectively.

Best practices for effective Chef implementation

To ensure a successful Chef implementation, it is important to follow best practices. These include modularizing cookbooks for reusability, using version control for cookbook management, leveraging testing frameworks for cookbook validation, implementing a staged rollout strategy for changes, and maintaining clear documentation.

Summary

In conclusion, automating infrastructure with Chef in Linux provides a powerful solution for managing and configuring complex environments. Chef's client-server architecture, declarative approach, and extensive feature set make it a popular choice for configuration management. By adopting Chef, organizations can achieve efficient infrastructure provisioning, consistent configurations, reduced errors, and enhanced security. While Chef has its challenges and limitations, it continues to evolve, and its vibrant community ensures ongoing development and support. As the demand for scalable and automated infrastructure grows, Chef remains a valuable tool for Linux-based systems.

In the next chapter, we will talk about best practices and security guidelines.

15

Security Guidelines and Best Practices

In the modern world, where almost all computers are connected to the internet and online applications play an increasingly larger role in all aspects of our lives, information security is also becoming more and more important. When information stored in digital form becomes more valuable and malicious actors constantly devise new attacks, every system administrator must make a conscious effort to keep their machines secure.

Luckily, following security guidelines and best practices can prevent most attacks and limit the impact of successful attacks if they occur.

In this chapter, we will learn about the following:

- Information security components and types of breaches
- Common types of attacks and threats
- Attack vectors and security vulnerabilities
- Ways to keep your system secure and stable

Common threats and attack types

There are many reasons why attackers may target a system, many ways to attack targets, and multiple possible consequences for the operator of the compromised system. Let's examine them in detail.

The motivation of attackers and the possible consequences

The picture of an attack on computer systems that movies, literature, and video games tend to show is usually an attack on a carefully selected target with a specific goal – most often, to steal some valuable information, modify it, or perhaps destroy it.

Such attacks certainly exist in the real world and they are a huge concern for high-profile companies and government agencies. However, that popular depiction often misleads people into believing that security is not important for them because they do not have any valuable information and are not high-profile targets.

That perception might have been correct in the early days of the internet but, these days, it is a very dangerous assumption. In reality, most attacks are no longer targeted and carefully prepared, but rather automated and opportunistic. Every machine connected to the internet is constantly probed by automated tools that attempt to exploit known security weaknesses.

Worse yet, automated attacks often rely on the availability of compromised third-party machines – running automated attack tools on machines that belong to the attackers themselves would be expensive and easy to detect and mitigate. To avoid paying for hosting and evade detection and blocking, attackers often gain unauthorized access to a few machines and use them to facilitate further attacks. A group of machines controlled by a malicious entity is often called a **botnet**. Botnets are used to probe more machines and take control of them, distribute malware, send unsolicited messages (spam), and perform other types of attacks.

Thus, a careless system administrator can become not only a victim of an attack but also an unintentional and unknowing accomplice of the attacker. In some cases, owners of compromised systems can come under investigation from law enforcement agencies and be suspected of conducting the attack because it came from their machines. Such cases are rare but even if the owner is not held legally liable, there are still many possible consequences of allowing an attacker to control your machine: cost of electricity (for on-premises machines or co-location) or CPU time on cloud platforms, bandwidth use fees, and system overload that takes resources away from legitimate users.

Finally, a machine that was identified as a source of attacks or spam can be added to blacklists. Multiple blacklists are maintained by various companies, so if your IP address or a domain name ends up in those lists, removing it from every list can be a very time-consuming endeavor. Moreover, blacklist maintainers are not required to remove your address since they are private companies, and blacklist inclusion and removal are not governed by any laws, and they may refuse to remove entries of repeat offenders or demand extensive proof that the current owner significantly improved its security practices.

Information security properties and attacks on them

The three components of information security are usually defined as follows:

- Availability
- Confidentiality
- Integrity (or authenticity)

Information availability means that authorized users can access it when they need it. Confidentiality means that users can only access the information they are authorized to access. Finally, integrity means that there are no accidental or deliberate modifications that are performed or authorized by legitimate users.

Attacks can have different goals – either to compromise information availability or to gain control of the target to compromise the confidentiality and authenticity of information stored on it. In the modern world, many attackers are also interested solely in the resources of the machine and not in any information stored on it. Let's look at these attacks in detail.

Denial of service

An attack on information availability is called a **Denial of Service** (**DoS**) attack. On the surface, it may look like the most benign type of attack because its effects are usually temporary. However, such attacks still can have severe consequences – for example, an online store whose website becomes unavailable can experience a significant loss of revenue, while an attack on a phone system may leave subscribers unable to make emergency calls and lead to loss of life as well. Some DoS attacks are performed simply as acts of vandalism but many such attacks are launched to extort money from the target system operator in exchange for stopping the attack, harm its reputation by rendering the service unreliable, or prevent it from making information available to users (the last goal is especially common for politically-motivated attacks).

There are two possible ways to perform a DoS attack. The classic way involves exploiting a flaw in the system software to crash it or make it repeatedly perform complex operations and thus slow it down. These attacks can be prevented by proper software development and configuration.

The other, increasingly more common, type is the **Distributed Denial of Service** (**DDoS**) attack. Such attacks use large numbers of machines to saturate the network link of the target system or overload it with requests beyond its capacity. The attacker can either generate attack traffic from a large botnet or use an amplification attack – that is, they can send DNS or NTP requests to public servers and specify the address of the attack target as the source address to make them send unsolicited reply packets to the target that the target never requested. Since replies are typically larger than requests, an amplification attack can save the attacker considerable bandwidth and computational resources by involving well-intentioned third parties in the attack.

The worst part of a DDoS attack is that if an attacker generates enough traffic to saturate the network link of the target, there is nothing the administrator of the target machine itself can do to mitigate it – if the attack traffic has reached the target, its damage is already done. Such attacks can only be mitigated by the hosting or internet service provider, or a dedicated DDoS protection service that filters out malicious packets and forwards legitimate requests to the target machine. However, DDoS attacks are always targeted and never opportunistic since they require sending traffic from multiple machines to a single designated target, and most systems never become DDoS attack targets.

Credential theft and brute-force attacks

Taking full control of a target machine is one of the most attractive goals for any attacker because it allows them to easily compromise the integrity and confidentiality of any information stored on it and use the machine itself for their purposes.

The cleanest way to gain access is to impersonate a legitimate user. If attackers somehow gain possession of a password, a cryptographic key, or an API key used for authentication, their use of the system will look indistinguishable from normal access.

Even credentials for access to small servers can be valuable in the modern world if the attacker can steal enough of them – one of the most common attacker actions after gaining access is to run cryptocurrency mining software on the compromised machine and thus directly convert its CPU and GPU power into money. Stealing credentials for access to cloud platforms, email services, or **Voice over Internet Protocol (VoIP)** is even more lucrative because attackers can use them to spawn new virtual machines, send spam, or make international calls – some even go as far as selling such illegally acquired resources to third parties who are unaware of their origin. The cost of those services, of course, has to be paid by the lawful owner of those credentials.

A lot of malware is programmed to steal passwords and keys from end user computers. This method is ideal for attackers because it leaves no traces on the target machines that they access using those stolen credentials.

However, many other attacks exploit the fact that end users often use weak passwords that are easy to guess. This is the basis for brute-force attacks, which are conducted by attempting to log in repeatedly with different passwords from a password dictionary that contains common words, commonly used passphrases, and often passwords stolen from other machines in the hope that someone used them for more than one machine or service (which is often the case).

Brute-force attacks can be made significantly harder to execute by using strong passwords, encrypted keys, and setting up rate limiting to give the attacker fewer chances to log in and guess the password.

Attacks using configuration and software vulnerabilities

Finally, in some cases, attackers can perform actions that logically must be denied to them by exploiting flaws in the system itself.

There are two classes of such flaws: configuration issues and software vulnerabilities.

For an example of a configuration flaw, consider that the standard protocol for email submission – **Simple Mail Transport Protocol (SMTP)** – does not require mandatory authentication. For this reason, every SMTP server implementation can be configured to allow anyone to send mail through it and act as an open relay. If a server with such a configuration is exposed to the public internet, attackers can use it to send large amounts of spam through it simply because it does not check whether the sender is a legitimate user of the system or not.

In other cases, the flaw is in the software itself. For a contrived example, suppose a web application implements user authentication and correctly redirects users to their account pages upon login – say, a user with a login name of `bob` gets redirected from `https://example.com/login` to `https://example.com/users/bob` when they enter their login and password. However, due to a programming mistake, the application never checks the user account when someone tries to access an account page, so anyone who knows that there is a user named `bob` in that system can access their account page simply by typing `https://example.com/users/bob` in the address bar.

This example may sound egregious but it is not very far from vulnerabilities sometimes found in real software, even if their exploitation method might be more complex than typing a URL into the address bar.

Luckily, most vulnerabilities are not as dangerous. When security researchers publish their findings and when software maintainers release fixes for discovered vulnerabilities, they use a set of terms for vulnerability types and severity levels that you should be familiar with to estimate how important a fix is.

Vulnerability reports are published by individual researchers and software vendors and also aggregated in databases such as the National Vulnerability Database, which is maintained by the United States **National Institute of Standards and Technology (NIST)**. Every known vulnerability in those databases is assigned a unique identifier such as CVE-2023-28531 (`https://nvd.nist.gov/vuln/detail/CVE-2023-28531`), where **CVE** stands for **Common Vulnerabilities and Exposures**.

The usual set of severity levels is as follows:

- **Critical**: Usually, this allows any attacker who can connect to the system to gain complete control of it. Vulnerable systems should be patched immediately or, if a patch is not available yet, isolated to make them inaccessible to attackers.

- **High and medium**: These may allow the attacker to significantly compromise the system but require special circumstances (for example, certain features enabled in the system) or difficult exploitation procedures. Affected systems should always be patched as soon as possible and may need temporary mitigation methods if a patch is not available (such as disabling the affected feature).

- **Low**: This does not give significant advantages to attackers as exploitation is only possible under rare circumstances.

Databases assign these levels based on multiple factors such as risk level, the skill required of an attacker to exploit the vulnerability, the impact, and more. For details, you may want to read about the Common Vulnerability Scoring System used by NIST or the OWASP Risk Rating Methodology.

Here are some common vulnerability types:

- Arbitrary code execution
- Privilege escalation
- Denial of service

Arbitrary code execution attacks tend to be the most dangerous because they allow attackers to introduce new program logic to the target system rather than merely use or abuse the software already running on it. However, there can be many mitigating factors. A vulnerability that allows remote unauthenticated attackers to execute arbitrary code by sending a specially crafted request to the target system over the network is the worst threat of all and may warrant taking the affected systems offline completely until they can be patched. However, if the user must be authenticated to execute the attack, it limits its impact greatly – to get to the arbitrary code execution stage, the attacker needs to steal valid user credentials first. Moreover, executing arbitrary code from a process that is running with highly restricted privileges and does not have access to any sensitive information may not benefit the attacker much.

Privilege escalation attacks allow legitimate users to perform actions that are not supposed to be available to them. Their impact can be severe – for example, if a vulnerability allows any authenticated user to read every file on the system regardless of its permissions, any machine where non-administrator users are allowed to log in is at risk of privacy breaches. However, before attackers outside of your organization can take advantage of it, they first need to find a way to log in. For systems where only administrators are allowed, such a vulnerability is not a concern at all, unless it is present at the same time with a remote arbitrary code execution vulnerability or credential theft.

Finally, **denial of service** vulnerabilities merely allow the attacker to compromise the availability of the system, as we already discussed.

With this in mind, let's discuss how to protect your system from those types of attacks.

Keeping your system secure

A common joke in information security circles is that the only perfectly secure system is one that is powered off. Such a system is only secure in the sense of integrity and confidentiality, of course – at the cost of availability. Any realistic scenario is always a compromise and there is always a risk; the system administrator's goal is to prevent known attacks and reduce the impact of unknown ones, and every administrator must always be ready to respond to new threats and mitigate them.

Luckily, following simple guidelines can considerably reduce the risk – let's discuss the general strategies and tactics to prevent specific attack types.

Reducing the attack surface

A system's attack surface is, roughly speaking, the set of all ways to access it. For example, a machine that is running a web server and also a mail server has a larger attack surface than a system that only runs one of those. If we assume that vulnerabilities and configuration issues are equally probable in those services and can arise independently, then a system running both is twice as likely to be vulnerable.

Of course, many real systems need to provide both services – for example, an email provider needs mail servers and a website for customers. Reducing the attack surface is not about running fewer services, but rather about isolating services from one another and, ideally, from attackers.

For example, a typical web application stack involves application servers and database servers. Most of the time, there is no reason for database servers to be publicly accessible – it is always a good idea to restrict access to them to the internal network. Moreover, application servers do not need to be *directly* publicly accessible either – as we discussed in *Chapter 13, High Availability*, they can be kept behind a load balancer. Not only will a load balancer improve the system's availability but it will also reduce the attack surfaces that are available to attackers from the public internet. Additionally, it can provide rate limiting and a threat detection system and shield application servers from at least some attack traffic.

Compartmentalization and privilege separation

A long time ago, the only way to isolate different components of a system from one another was to run them on different physical machines. This approach is very expensive and is only practiced when there are also significant reasons to do so – for example, applications designed to handle high loads have to use separate databases and application servers simply to meet their performance requirements, rather than solely to reduce the attack surface of the system.

However, in the last two decades, there have been many more granular methods to isolate processes, even on commodity hardware. Virtualization allows you to run multiple instances of operating systems on a single physical machine, and modern virtual machine managers make it easy to spawn them – not to mention cloud platforms that allow VMs to be spawned with your operating system of choice with a single click or API call.

Apart from full virtualization, there are many ways to isolate applications on a single machine from one another. Those include `chroot` environments, containers, and mandatory access control systems.

Using chroot

The oldest way to completely separate processes is `chroot`. Technically, `chroot` (change root) is a system call in the kernel that changes the root directory for a process. A process whose root directory was changed no longer has access to any files outside of it – to that process, it looks as if its directory is all that exists in the system. Setting up a `chroot` environment for a process by hand may be a time-consuming task, so many distributions provide special packages to simplify the process.

For example, Fedora provides an easy way to run ISC BIND (also known as `named`), a popular DNS server, in `chroot`:

```
$ sudo dnf install bind9-next-chroot
Installed:
  bind9-next-32:9.19.11-1.fc37.x86_64      bind9-next-
chroot-32:9.19.11-1.fc37.x86_64      bind9-next-dnssec-
```

```
utils-32:9.19.11-1.fc37.x86_64        bind9-next-libs-32:9.19.11-1.fc37.
x86_64      bind9-next-license-32:9.19.11-1.fc37.noarch
Complete!
$ sudo systemctl enable  named-chroot
Created symlink /etc/systemd/system/multi-user.target.wants/named-
chroot.service → /usr/lib/systemd/system/named-chroot.service.
$ tree /var/named/chroot/
/var/named/chroot/
├── dev
├── etc
│   ├── crypto-policies
│   │   └── back-ends
│   ├── named
│   └── pki
│       └── dnssec-keys
├── proc
│   └── sys
│       └── net
│           └── ipv4
...
```

As you can see, the directory that will serve as a limited root for the BIND process mimics the real root in that it also has /dev and /proc hierarchies – global versions of those directories become unavailable to the process.

Thus, even if attackers manage to inject malicious code into the BIND process using a remote code execution vulnerability, they will be unable to read any files outside /var/named/chroot. However, they will still be free to interact with other processes on the system. If deeper isolation is required, you can use containers instead.

Using containers

The Linux kernel provides a container technology named **LXC**. It consists of multiple sub-components such as process groups, control groups, and network namespaces. Creating container environments and launching them by hand is a laborious process, so people created tools that automated the process, as well as registries of ready-to-use container images.

At this time, the most popular tool for managing containers on Linux is Docker, although there are alternatives to it, such as Podman. We will demonstrate process isolation by launching a Fedora image using Docker:

```
$ sudo dnf install docker
$ sudo docket pull fedora:latest
$ docker run -it fedora:latest bash
[root@df231cc10b87 /]# dnf install procps-ng
```

```
[root@df231cc10b87 /]# ps aux
USER            PID %CPU %MEM     VSZ     RSS TTY       STAT START     TIME
COMMAND
root              1  0.0  0.0    4720    3712 pts/0     Ss   15:21    0:00 /
bin/bash
root             55  0.0  0.0    5856    2688 pts/0     R+   15:23    0:00 ps
aux
```

As you can see, when we start a bash shell process in a container, from inside the container, it looks as if it is the only process in the system. None of the processes of the host system, such as systemd or the Docker daemon process, are visible and processes inside the container cannot interact with them in any way.

Using mandatory access control

Finally, there are ways to grant users and processes different capabilities – ideally, only the capabilities they need to function. Such mechanisms are known as **Mandatory Access Control** (**MAC**) systems – as opposed to the classic **Discretionary Access Control** – that is, the Unix file permission system.

The most popular MAC system for Linux these days is **Security Enhanced Linux** (**SELinux**), although there are less common alternatives as well – most notably, the AppArmor kernel module.

These days, many Linux distributions enable SELinux by default and include capabilities and object contexts for popular services.

You can check whether SELinux is enabled on your system with the getenforce command:

```
$ getenforce
Enforcing
```

In Enforcing mode, SELinux disallows actions not permitted by capabilities granted to a user or a process – for example, a process without the can_network_connect capability will not be able to initiate any network connections. In Permissive mode, SELinux will generate alerts in the system log but not enforce capabilities – this mode is ideal for testing. Finally, if the mode is Disabled, SELinux does not perform any capability checks at all.

Configuring SELinux policies is a subject far beyond the scope of an introductory book. However, if you choose to keep SELinux enabled, you will often need to know how to grant processes access to files.

For example, in the Apache HTTPd configuration files, it is possible to specify any directory as a website root. That is also a possible security problem if an attacker gains control of the Apache process. That is why, in Fedora, the maintainers of that package introduced a set of special SELinux contexts for files that must be accessible to the web server.

You can view the SELinux contexts of a file by adding the -Z option to the ls command. For example, the /var/lib/httpd directory has the httpd_var_lib_t context, which grants read access:

```
$ sudo ls -alZ /var/lib/httpd/
total 8
drwx------.  2 apache apache system_u:object_r:httpd_var_lib_t:s0 4096
Jun 17  2022 .
drwxr-xr-x. 55 root    root   system_u:object_r:var_lib_t:s0         4096
Nov 16 02:39 ..
```

Website directories that may be writable use a different context – httpd_sys_content_t:

```
$ ls -alZ /var/www/html/
total 8
drwxr-xr-x. 2 root root system_u:object_r:httpd_sys_content_t:s0 4096
Jun 17  2022 .
drwxr-xr-x. 6 root root system_u:object_r:httpd_sys_content_t:s0 4096
Jun 17  2022 ..
```

If you set a website root to a newly created directory elsewhere, the Apache process will not have access to it because it lacks the required context, even if it should be readable to it according to Unix permissions. You can grant it access using the chcon command:

```
$ sudo chcon system_u:object_r:httpd_sys_content_t:s0 /home/webmaster/
public_html
```

There are many more possibilities for creating flexible and fine-grained security policies with SELinux, but beginners should certainly start by using the distribution defaults first.

Preventing credential theft and brute-force attacks

Credential theft must be addressed at user workstations. There is no definitive way to prevent credential theft, so you should strive to protect your laptop or desktop system from attacks in general – keep it up to date, protect it from malware, and avoid falling for social engineering attacks such as phishing letters that contain malicious links.

When it comes to brute-force attacks, there are two complementary approaches – keeping passwords hard to guess and limiting authentication attempts.

Most Linux distributions have a PAM module named pam_pwquality.so enabled by default, which prevents unprivileged users from using insecure passwords. You can verify this by trying to set your password to something way too short or a simple dictionary word:

```
$ passwd
Changing password for user ...
Current password:
New password: qwerty
```

```
BAD PASSWORD: The password is shorter than 8 characters
passwd: Authentication token manipulation error
$ passwd
Changing password for user ...
Current password:
New password: swordfish
BAD PASSWORD: The password fails the dictionary check - it is based on
a dictionary word
passwd: Authentication token manipulation error
```

The backend that it uses for password strength checking is called `cracklib`. It normally keeps its data in `/usr/share/cracklib`. Its dictionary is in a binary format but it offers tools for manipulating those dictionary files and checking password strength without actually trying to set a password:

```
$ echo "swordfish" | cracklib-check
swordfish: it is based on a dictionary word
```

Note that the root user is exempt from password quality checking and is free to set any password for any user.

There is also a tool for generating secure random passwords named **pwgen** that is present in the package repositories of most Linux distributions. It allows you to specify whether passwords should be completely random, what characters they should contain, how long they should be, and how many passwords to generate at once:

```
$ pwgen --secure 40 1
DMYRsJNuXJQb98scCUASDzt3GFPa7yzGg9eS1L1U
$ pwgen --secure 8 16
Ob4r24Sp KwGCF63L 3bgKI79L aWK2K7aK MFf1ykum y74VTKqb OxbrNlI0
8Dl4yilz
bmI1RsKr oM4p8dsJ 2EXmmHIJ bt9Go1gg 3znmVqpo vZ9BDI1z QMZ7eME0
izkB15Xe
```

Rate limiting is a broad subject and its setup varies between services and applications. However, there are also general integrated solutions that support multiple services, such as **fail2ban**.

Reducing the risk of software vulnerabilities

The best way to ensure that your system does not have vulnerable software in it is to use software from your distribution repositories and install updates on time. Installing packages by hand should be avoided because such packages will not automatically receive updates from the distribution maintainers. If they are required, you should make sure you subscribe to their release announcements and check for updates yourself.

A lot of the time, vulnerabilities are found in already released software versions rather than during development – this situation is called a **zero-day vulnerability** (often shortened to **0day**). Sometimes, they turn out to have existed for months or even years before their discovery. Updated packages that fix such vulnerabilities may appear later than malicious actors start exploiting them. In that case, software maintainers often suggest a temporary mitigation strategy that may involve changing configuration options, disabling specific features, or making other changes in the system to make the attack harder to execute. For this reason, it is also a good idea to follow the blogs or mailing lists or the distribution and software projects that you use often.

Summary

In this chapter, we learned about various types of attacks on computer systems, attackers' motivation to execute them, and possible consequences for users.

We also learned about common strategies for keeping your systems protected from attacks. However, we only scratched the surface – information security is a large field, and keeping your knowledge of it up to date and your systems safe is a life-long pursuit for every system administrator.

Further reading

To learn more about the topics that were covered in this chapter, take a look at the following resources:

- NIST vulnerability database: `https://nvd.nist.gov/`
- Common Vulnerability Scoring System: `https://nvd.nist.gov/vuln-metrics/cvss`
- OWASP Risk Rating Methodology: `https://owasp.org/www-community/OWASP_Risk_Rating_Methodology`
- Fail2Ban rate limiting daemon: `https://www.fail2ban.org`

Index

C

Packtpub.com

Subscribe to our online digital library for full access to over 7,000 books and videos, as well as industry leading tools to help you plan your personal development and advance your career. For more information, please visit our website.

Why subscribe?

- Spend less time learning and more time coding with practical eBooks and Videos from over 4,000 industry professionals

- Improve your learning with Skill Plans built especially for you

- Get a free eBook or video every month

- Fully searchable for easy access to vital information

- Copy and paste, print, and bookmark content

Did you know that Packt offers eBook versions of every book published, with PDF and ePub files available? You can upgrade to the eBook version at packtpub.com and as a print book customer, you are entitled to a discount on the eBook copy. Get in touch with us at customercare@packtpub.com for more details.

At www.packtpub.com, you can also read a collection of free technical articles, sign up for a range of free newsletters, and receive exclusive discounts and offers on Packt books and eBooks.

Other Books You May Enjoy

If you enjoyed this book, you may be interested in these other books by Packt:

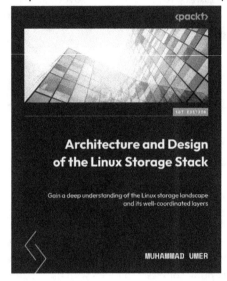

Architecture and Design of the Linux Storage Stack

Muhammad Umer

ISBN: 978-1-83763-996-0

- Understand the role of the virtual filesystem
- Explore the different flavors of Linux filesystems and their key concepts
- Manage I/O operations to and from block devices using the block layer
- Deep dive into the Small Computer System Interface (SCSI) subsystem and the layout of physical devices
- Gauge I/O performance at each layer of the storage stack
- Discover the best storage practices

Mastering Linux Security and Hardening - Third Edition

Donald A. Tevault

ISBN: 978-1-83763-051-6

- Prevent malicious actors from compromising a production Linux system

- Leverage additional features and capabilities of Linux in this new version

- Use locked-down home directories and strong passwords to create user accounts

- Prevent unauthorized people from breaking into a Linux system

- Configure file and directory permissions to protect sensitive data

- Harden the Secure Shell service in order to prevent break-ins and data loss

- Apply security templates and set up auditing

Packt is searching for authors like you

If you're interested in becoming an author for Packt, please visit `authors.packtpub.com` and apply today. We have worked with thousands of developers and tech professionals, just like you, to help them share their insight with the global tech community. You can make a general application, apply for a specific hot topic that we are recruiting an author for, or submit your own idea.

Share Your Thoughts

Now you've finished *Linux for System Administrators*, we'd love to hear your thoughts! Scan the QR code below to go straight to the Amazon review page for this book and share your feedback or leave a review on the site that you purchased it from.

`https://packt.link/r/1803247940`

Your review is important to us and the tech community and will help us make sure we're delivering excellent quality content.

Download a free PDF copy of this book

Thanks for purchasing this book!

Do you like to read on the go but are unable to carry your print books everywhere?

Is your eBook purchase not compatible with the device of your choice?

Don't worry, now with every Packt book you get a DRM-free PDF version of that book at no cost.

Read anywhere, any place, on any device. Search, copy, and paste code from your favorite technical books directly into your application.

The perks don't stop there, you can get exclusive access to discounts, newsletters, and great free content in your inbox daily

Follow these simple steps to get the benefits:

1. Scan the QR code or visit the link below

https://packt.link/free-ebook/9781803247946

2. Submit your proof of purchase
3. That's it! We'll send your free PDF and other benefits to your email directly